T0291504

LONDON MATHEMATICAL SOCIETY LECTURE NOTE SERIES

Managing Editor: Professor M. Reid, Mathematics Institute, University of Warwick, Coventry CV4 7AL, United Kingdom

The titles below are available from booksellers, or from Cambridge University Press at www.cambridge.org/mathematics

225 A mathematical introduction to string theory, S. ALBEVERIO *et al*
226 Novikov conjectures, index theorems and rigidity I, S.C. FERRY, A. RANICKI & J. ROSENBERG (eds)
227 Novikov conjectures, index theorems and rigidity II, S.C. FERRY, A. RANICKI & J. ROSENBERG (eds)
228 Ergodic theory of $\mathbf{Z}^d$-actions, M. POLLICOTT & K. SCHMIDT (eds)
229 Ergodicity for infinite dimensional systems, G. DA PRATO & J. ZABCZYK
230 Prolegomena to a middlebrow arithmetic of curves of genus 2, J.W.S. CASSELS & E.V. FLYNN
231 Semigroup theory and its applications, K.H. HOFMANN & M.W. MISLOVE (eds)
232 The descriptive set theory of Polish group actions, H. BECKER & A.S. KECHRIS
233 Finite fields and applications, S. COHEN & H. NIEDERREITER (eds)
234 Introduction to subfactors, V. JONES & V.S. SUNDER
235 Number theory: Séminaire de théorie des nombres de Paris 1993–94, S. DAVID (ed)
236 The James forest, H. FETTER & B. GAMBOA DE BUEN
237 Sieve methods, exponential sums, and their applications in number theory, G.R.H. GREAVES *et al* (eds)
238 Representation theory and algebraic geometry, A. MARTSINKOVSKY & G. TODOROV (eds)
240 Stable groups, F.O. WAGNER
241 Surveys in combinatorics, 1997, R.A. BAILEY (ed)
242 Geometric Galois actions I, L. SCHNEPS & P. LOCHAK (eds)
243 Geometric Galois actions II, L. SCHNEPS & P. LOCHAK (eds)
244 Model theory of groups and automorphism groups, D.M. EVANS (ed)
245 Geometry, combinatorial designs and related structures, J.W.P. HIRSCHFELD *et al* (eds)
246 p-Automorphisms of finite p-groups, E.I. KHUKHRO
247 Analytic number theory, Y. MOTOHASHI (ed)
248 Tame topology and O-minimal structures, L. VAN DEN DRIES
249 The atlas of finite groups - Ten years on, R.T. CURTIS & R.A. WILSON (eds)
250 Characters and blocks of finite groups, G. NAVARRO
251 Gröbner bases and applications, B. BUCHBERGER & F. WINKLER (eds)
252 Geometry and cohomology in group theory, P.H. KROPHOLLER, G.A. NIBLO & R. STÖHR (eds)
253 The q-Schur algebra, S. DONKIN
254 Galois representations in arithmetic algebraic geometry, A.J. SCHOLL & R.L. TAYLOR (eds)
255 Symmetries and integrability of difference equations, P.A. CLARKSON & F.W. NIJHOFF (eds)
256 Aspects of Galois theory, H. VÖLKLEIN, J.G. THOMPSON, D. HARBATER & P. MÜLLER (eds)
257 An introduction to noncommutative differential geometry and its physical applications (2nd edition), J. MADORE
258 Sets and proofs, S.B. COOPER & J.K. TRUSS (eds)
259 Models and computability, S.B. COOPER & J. TRUSS (eds)
260 Groups St Andrews 1997 in Bath I, C.M. CAMPBELL *et al* (eds)
261 Groups St Andrews 1997 in Bath II, C.M. CAMPBELL *et al* (eds)
262 Analysis and logic, C.W. HENSON, J. IOVINO, A.S. KECHRIS & E. ODELL
263 Singularity theory, W. BRUCE & D. MOND (eds)
264 New trends in algebraic geometry, K. HULEK, F. CATANESE, C. PETERS & M. REID (eds)
265 Elliptic curves in cryptography, I. BLAKE, G. SEROUSSI & N. SMART
267 Surveys in combinatorics, 1999, J.D. LAMB & D.A. PREECE (eds)
268 Spectral asymptotics in the semi-classical limit, M. DIMASSI & J. SJÖSTRAND
269 Ergodic theory and topological dynamics of group actions on homogeneous spaces, M.B. BEKKA & M. MAYER
271 Singular perturbations of differential operators, S. ALBEVERIO & P. KURASOV
272 Character theory for the odd order theorem, T. PETERFALVI. Translated by R. SANDLING
273 Spectral theory and geometry, E.B. DAVIES & Y. SAFAROV (eds)
274 The Mandelbrot set, theme and variations, T. LEI (ed)
275 Descriptive set theory and dynamical systems, M. FOREMAN, A.S. KECHRIS, A. LOUVEAU & B. WEISS (eds)
276 Singularities of plane curves, E. CASAS-ALVERO
277 Computational and geometric aspects of modern algebra, M. ATKINSON *et al* (eds)
278 Global attractors in abstract parabolic problems, J.W. CHOLEWA & T. DLOTKO
279 Topics in symbolic dynamics and applications, F. BLANCHARD, A. MAASS & A. NOGUEIRA (eds)
280 Characters and automorphism groups of compact Riemann surfaces, T. BREUER
281 Explicit birational geometry of 3-folds, A. CORTI & M. REID (eds)
282 Auslander–Buchweitz approximations of equivariant modules, M. HASHIMOTO
283 Nonlinear elasticity, Y.B. FU & R.W. OGDEN (eds)
284 Foundations of computational mathematics, R. DEVORE, A. ISERLES & E. SÜLI (eds)
285 Rational points on curves over finite fields, H. NIEDERREITER & C. XING
286 Clifford algebras and spinors (2nd Edition), P. LOUNESTO
287 Topics on Riemann surfaces and Fuchsian groups, E. BUJALANCE, A.F. COSTA & E. MARTÍNEZ (eds)
288 Surveys in combinatorics, 2001, J.W.P. HIRSCHFELD (ed)
289 Aspects of Sobolev-type inequalities, L. SALOFF-COSTE
290 Quantum groups and Lie theory, A. PRESSLEY (ed)
291 Tits buildings and the model theory of groups, K. TENT (ed)
292 A quantum groups primer, S. MAJID
293 Second order partial differential equations in Hilbert spaces, G. DA PRATO & J. ZABCZYK
294 Introduction to operator space theory, G. PISIER
295 Geometry and integrability, L. MASON & Y. NUTKU (eds)
296 Lectures on invariant theory, I. DOLGACHEV
297 The homotopy category of simply connected 4-manifolds, H.-J. BAUES
298 Higher operads, higher categories, T. LEINSTER (ed)
299 Kleinian groups and hyperbolic 3-manifolds, Y. KOMORI, V. MARKOVIC & C. SERIES (eds)
300 Introduction to Möbius differential geometry, U. HERTRICH-JEROMIN
301 Stable modules and the D(2)-problem, F.E.A. JOHNSON
302 Discrete and continuous nonlinear Schrödinger systems, M.J. ABLOWITZ, B. PRINARI & A.D. TRUBATCH
303 Number theory and algebraic geometry, M. REID & A. SKOROBOGATOV (eds)
304 Groups St Andrews 2001 in Oxford I, C.M. CAMPBELL, E.F. ROBERTSON & G.C. SMITH (eds)
305 Groups St Andrews 2001 in Oxford II, C.M. CAMPBELL, E.F. ROBERTSON & G.C. SMITH (eds)

306 Geometric mechanics and symmetry, J. MONTALDI & T. RATIU (eds)
307 Surveys in combinatorics 2003, C.D. WENSLEY (ed.)
308 Topology, geometry and quantum field theory, U.L. TILLMANN (ed)
309 Corings and comodules, T. BRZEZINSKI & R. WISBAUER
310 Topics in dynamics and ergodic theory, S. BEZUGLYI & S. KOLYADA (eds)
311 Groups: topological, combinatorial and arithmetic aspects, T.W. MÜLLER (ed)
312 Foundations of computational mathematics, Minneapolis 2002, F. CUCKER *et al* (eds)
313 Transcendental aspects of algebraic cycles, S. MÜLLER-STACH & C. PETERS (eds)
314 Spectral generalizations of line graphs, D. CVETKOVIĆ, P. ROWLINSON & S. SIMIĆ
315 Structured ring spectra, A. BAKER & B. RICHTER (eds)
316 Linear logic in computer science, T. EHRHARD, P. RUET, J.-Y. GIRARD & P. SCOTT (eds)
317 Advances in elliptic curve cryptography, I.F. BLAKE, G. SEROUSSI & N.P. SMART (eds)
318 Perturbation of the boundary in boundary-value problems of partial differential equations, D. HENRY
319 Double affine Hecke algebras, I. CHEREDNIK
320 L-functions and Galois representations, D. BURNS, K. BUZZARD & J. NEKOVÁR (eds)
321 Surveys in modern mathematics, V. PRASOLOV & Y. ILYASHENKO (eds)
322 Recent perspectives in random matrix theory and number theory, F. MEZZADRI & N.C. SNAITH (eds)
323 Poisson geometry, deformation quantisation and group representations, S. GUTT *et al* (eds)
324 Singularities and computer algebra, C. LOSSEN & G. PFISTER (eds)
325 Lectures on the Ricci flow, P. TOPPING
326 Modular representations of finite groups of Lie type, J.E. HUMPHREYS
327 Surveys in combinatorics 2005, B.S. WEBB (ed)
328 Fundamentals of hyperbolic manifolds, R. CANARY, D. EPSTEIN & A. MARDEN (eds)
329 Spaces of Kleinian groups, Y. MINSKY, M. SAKUMA & C. SERIES (eds)
330 Noncommutative localization in algebra and topology, A. RANICKI (ed)
331 Foundations of computational mathematics, Santander 2005, L.M PARDO, A. PINKUS, E. SÜLI & M.J. TODD (eds)
332 Handbook of tilting theory, L. ANGELERI HÜGEL, D. HAPPEL & H. KRAUSE (eds)
333 Synthetic differential geometry (2nd Edition), A. KOCK
334 The Navier–Stokes equations, N. RILEY & P. DRAZIN
335 Lectures on the combinatorics of free probability, A. NICA & R. SPEICHER
336 Integral closure of ideals, rings, and modules, I. SWANSON & C. HUNEKE
337 Methods in Banach space theory, J.M.F. CASTILLO & W.B. JOHNSON (eds)
338 Surveys in geometry and number theory, N. YOUNG (ed)
339 Groups St Andrews 2005 I, C.M. CAMPBELL, M.R. QUICK, E.F. ROBERTSON & G.C. SMITH (eds)
340 Groups St Andrews 2005 II, C.M. CAMPBELL, M.R. QUICK, E.F. ROBERTSON & G.C. SMITH (eds)
341 Ranks of elliptic curves and random matrix theory, J.B. CONREY, D.W. FARMER, F. MEZZADRI & N.C. SNAITH (eds)
342 Elliptic cohomology, H.R. MILLER & D.C. RAVENEL (eds)
343 Algebraic cycles and motives I, J. NAGEL & C. PETERS (eds)
344 Algebraic cycles and motives II, J. NAGEL & C. PETERS (eds)
345 Algebraic and analytic geometry, A. NEEMAN
346 Surveys in combinatorics 2007, A. HILTON & J. TALBOT (eds)
347 Surveys in contemporary mathematics, N. YOUNG & Y. CHOI (eds)
348 Transcendental dynamics and complex analysis, P.J. RIPPON & G.M. STALLARD (eds)
349 Model theory with applications to algebra and analysis I, Z. CHATZIDAKIS, D. MACPHERSON, A. PILLAY & A. WILKIE (eds)
350 Model theory with applications to algebra and analysis II, Z. CHATZIDAKIS, D. MACPHERSON, A. PILLAY & A. WILKIE (eds)
351 Finite von Neumann algebras and masas, A.M. SINCLAIR & R.R. SMITH
352 Number theory and polynomials, J. MCKEE & C. SMYTH (eds)
353 Trends in stochastic analysis, J. BLATH, P. MÖRTERS & M. SCHEUTZOW (eds)
354 Groups and analysis, K. TENT (ed)
355 Non-equilibrium statistical mechanics and turbulence, J. CARDY, G. FALKOVICH & K. GAWEDZKI
356 Elliptic curves and big Galois representations, D. DELBOURGO
357 Algebraic theory of differential equations, M.A.H. MACCALLUM & A.V. MIKHAILOV (eds)
358 Geometric and cohomological methods in group theory, M.R. BRIDSON, P.H. KROPHOLLER & I.J. LEARY (eds)
359 Moduli spaces and vector bundles, L. BRAMBILA-PAZ, S.B. BRADLOW, O. GARCÍA-PRADA & S. RAMANAN (eds)
360 Zariski geometries, B. ZILBER
361 Words: Notes on verbal width in groups, D. SEGAL
362 Differential tensor algebras and their module categories, R. BAUTISTA, L. SALMERÓN & R. ZUAZUA
363 Foundations of computational mathematics, Hong Kong 2008, F. CUCKER, A. PINKUS & M.J. TODD (eds)
364 Partial differential equations and fluid mechanics, J.C. ROBINSON & J.L. RODRIGO (eds)
365 Surveys in combinatorics 2009, S. HUCZYNSKA, J.D. MITCHELL & C.M. RONEY-DOUGAL (eds)
366 Highly oscillatory problems, B. ENGQUIST, A. FOKAS, E. HAIRER & A. ISERLES (eds)
367 Random matrices: High dimensional phenomena, G. BLOWER
368 Geometry of Riemann surfaces, F.P. GARDINER, G. GONZÁLEZ-DIEZ & C. KOUROUNIOTIS (eds)
369 Epidemics and rumours in complex networks, M. DRAIEF & L. MASSOULIÉ
370 Theory of *p*-adic distributions, S. ALBEVERIO, A.YU. KHRENNIKOV & V.M. SHELKOVICH
371 Conformal fractals, F. PRZYTYCKI & M. URBAŃSKI
372 Moonshine: The first quarter century and beyond, J. LEPOWSKY, J. MCKAY & M.P. TUITE (eds)
373 Smoothness, regularity and complete intersection, J. MAJADAS & A. G. RODICIO
374 Geometric analysis of hyperbolic differential equations: An introduction, S. ALINHAC
375 Triangulated categories, T. HOLM, P. JØRGENSEN & R. ROUQUIER (eds)
376 Permutation patterns, S. LINTON, N. RUŠKUC & V. VATTER (eds)
377 An introduction to Galois cohomology and its applications, G. BERHUY
378 Probability and mathematical genetics, N.H. BINGHAM & C.M. GOLDIE (eds)
379 Finite and algorithmic model theory, J. ESPARZA, C. MICHAUX & C. STEINHORN (eds)
380 Real and complex singularities, M. MANOEL, M.C. ROMERO FUSTER & C.T.C WALL (eds)
381 Symmetries and integrability of difference equations, D. LEVI, P. OLVER, Z. THOMOVA & P. WINTERNITZ (eds)

London Mathematical Society Lecture Note Series: 382

Forcing with Random Variables and Proof Complexity

JAN KRAJÍČEK
Charles University, Prague

CAMBRIDGE
UNIVERSITY PRESS

University Printing House, Cambridge CB2 8BS, United Kingdom

One Liberty Plaza, 20th Floor, New York, NY 10006, USA

477 Williamstown Road, Port Melbourne, VIC 3207, Australia

314-321, 3rd Floor, Plot 3, Splendor Forum, Jasola District Centre, New Delhi - 110025, India

103 Penang Road, #05-06/07, Visioncrest Commercial, Singapore 238467

Cambridge University Press is part of the University of Cambridge.

It furthers the University's mission by disseminating knowledge in the pursuit of education, learning and research at the highest international levels of excellence.

www.cambridge.org
Information on this title: www.cambridge.org/9780521154338

First published 2011

A catalogue record for this publication is available from the British Library

Library of Congress Cataloging in Publication data
Krajíček, Jan
Forcing with random variables and proof complexity / Jan Krajíček.
p. cm. – (London Mathematical Society Lecture Note Series: 382)
Includes bibliographical references and indexes.
ISBN 978-0-521-15433-8 (pbk.)
1. Computational complexity. 2. Random variables. 3. Mathematical analysis.
I. Title. II. Series.
QA267.7.K73 2011
511.3′6–dc22 2010036194

ISBN 978-0-521-15433-8 Paperback

To my parents

Contents

Preface

Proof complexity is concerned with the mathematical analysis of the informal concept of a feasible proof when the qualification 'feasible' is interpreted in a complexity-theoretic sense. The most important measure of complexity of a proof is its length when it is thought of as a string over a finite alphabet. The basic question that proof complexity studies is to estimate (from below as well as from above) the minimal possible length of a proof of a formula. Measuring the complexity of a proof by its length may seem crude at first but it is analogous to measuring the complexity of an algorithm by the length of time it takes.

In the context of propositional logic the main question is whether there exists a proof system in which every propositional tautology has a short proof, a proof bounded in length by a polynomial in the length of the formula. With a suitably general definition of what a 'proof system' is, the question is equivalent to the problem whether the computational complexity class NP is closed under complementation.

In the setting of first-order logic one considers theories whose principal axiom scheme is the scheme of induction but accepted only for predicates on binary strings that have limited computational complexity. These are the so-called bounded arithmetic theories. A typical question is this: Can we prove more universally valid properties of strings if we assume induction for NP predicates than if we have only induction for polynomial-time predicates?

These two strands of proof complexity are, in fact, very much bound together and in a precise technical sense one can think of proof systems as non-uniform versions of theories. The two problems mentioned, and their variants, are also quite linked with fundamental problems of computational complexity theory such as the P versus NP problem, the existence of one-way functions, or the possibility of a universal derandomization.

Answering the propositional problem in the negative, that is, showing that NP is not closed under complementation, implies of course that P and NP are different. Answering the first-order question in the negative would not necessarily separate P and NP but it would show (among other things) that the conjecture that P differs from NP is consistent with Cook's theory PV (standing for polynomially verifiable). PV has names for all polynomial time algorithms and axioms codifying how these algorithms are built from each other. This would be quite significant as most of contemporary computational complexity theory can be formalized in PV or in its mild extensions.

Interestingly, it is known that the consistency of the statement that P differs from NP with PV follows from a super-polynomial lower bound on the length of Extended Frege propositional proofs of *any* sequence of tautologies. In fact, the consistency of the conjecture $P \neq NP$ with an arbitrary theory T (axiomatized by schemes) follows from any super-polynomial lower bound for a specific propositional proof system P depending on T. This is perhaps one of the reasons why proving lengths-of-proofs lower bounds appears quite difficult even for seemingly simple proof systems.

Obviously, when studying a theory it is quite useful to have a rich class of models. In particular, a separation of two theories may be proved by exhibiting a suitable model. However, it is also known that proving lengths-of-proofs lower bounds for propositional proof systems is equivalent to constructing extensions of particular bounded arithmetic models. For this reason we are interested in methods for constructing models of various bounded arithmetic theories.

In these notes we describe a new class of models of bounded arithmetic. The models are Boolean-valued and are based on random variables. We suggest that the body of results obtained in these notes shows that this provides a coherent and rich framework for studying bounded arithmetic and propositional proof complexity. In particular, we propose this as a framework in which it is possible to think about unconditional independence results for bounded arithmetic theories and about lengths-of-proofs lower bounds for strong proof systems.

Acknowledgements

I thank S. A. Cook (Toronto), E. Jeřábek (Prague), P. Pudlák (Prague), L. van den Dries (Urbana), S. Todorcevic (Paris/Toronto) and I. Tzameret (Prague) for discussions of related topics and for comments on drafts of parts of the book. I am especially indebted to P. Nguyen (Prague/Montreal) and N. Thapen (Prague) for extensive comments on the draft of the first seven parts, and to S. A. Cook for enduring alone my presentation of attempts at the construction of a model aimed at in Chapter 22 in May 2006 during the program *Logic and Algorithms* at the Isaac Newton Institute in Cambridge.

I lectured about parts of this work at seminars in Oxford in May and November 2004, at the Institute for Advanced Study in Princeton in February 2005, at the Isaac Newton Institute in Cambridge in April 2006, and at the meeting 'Computability in Europe' in Swansea in July 2006. I also gave a semester-long exposition of parts of this work at the Charles University in Prague in Fall 2004. I thank participants of all these events for a number of comments.

Support acknowledgments

I started to work on this book while I was a member of the Institute for Advanced Study in Princeton in Spring 2004, supported by the NSF grant DMS-0111298. Parts of this work were done while visiting the Mathematical Institute in Oxford supported by the EPSRC grant GR/S16997/01, the Institute for Advanced Study in Princeton supported by the NSF grant DMS-0111298, and the Isaac Newton Institute in Cambridge (program Logic and Algorithms) supported by an EPSRC grant N09176. This work has also been partially supported over the years by various Czech grant agencies: grants A1019401, IAA100190902 and AV0Z10190503 (AS CR), grant MSM0021620839, and grant LC505 (Eduard

Čech Center). A grant from the John F. Templeton Foundation provided partial support during 2008–2010.

For a large part of the work on this project, until Spring 2009, I have been a member of the Institute of Mathematics of the Academy of Sciences of the Czech Republic in Prague.

Introduction

Propositional proof complexity studies the lengths of propositional proofs or equivalently the time complexity of non-deterministic algorithms accepting some coNP-complete set. The main problem is the NP versus coNP problem, a question whether the computational complexity class NP is closed under complementation. Central objects studied are propositional proof systems (non-deterministic algorithms accepting the set of propositional tautologies). Time lower bounds then correspond to lengths-of-proofs lower bounds.

Bounded arithmetic is a generic name for a collection of first-order and second-order theories of arithmetic linked to propositional proof systems (and to a variety of other computational complexity topics). The qualification *bounded* refers to the fact that the induction axiom is typically restricted to a subclass of bounded formulas.

The links between propositional proof systems and bounded arithmetic theories have many facets but informally one can view them as two sides of the same thing: the former is a non-uniform version of the latter. In particular, it is known that proving lengths-of-proofs lower bounds for propositional proof systems is very much related to proving independence results for bounded arithmetic theories. In fact, proving such lower bounds is *equivalent* to constructing non-elementary extensions of particular models of bounded arithmetic theories. This offers a very clean and coherent framework for thinking about lengths-of-proofs lower bounds, one that has been quite successful in the past (let us mention just Ajtai's [2] lower bound for the lengths of proofs of the pigeonhole principle in constant-depth Frege systems, see Chapter 21).

In this book we introduce a new method for constructing bounded arithmetic models, and hence for proving independence results and lengths-of-proofs lower bounds. A brief description could be *forcing with random variables* but it also has features of non-standard analysis and of definable ultraproducts. The novelty lies neither in using forcing in bounded arithmetic or proof complexity (see 'Remarks on the literature' below) nor in forcing with random variables (that is well-established in set theory; see Scott [100] or Jech[49]), but rather in finding a way how to do this meaningfully in arithmetic, and further in using families of random variables that are sampled by algorithms restricted in a particular way (different from one application to another) rather than using the family of all random variables with a given sample space and range.

The models are built from random variables defined on a sample space Ω which is a non-standard finite set (often parameterized by a subset of $\{0,1\}^n$ with a non-standard n), and sampled by functions of some restricted computational complexity. This is considered inside an $\aleph_1$-saturated non-standard model of true arithmetic. One could equivalently work with sequences of bigger and bigger sample spaces and random variables defined on them, and consider their limit behavior using a suitable ultrafilter on $\mathbf{N}$, simulating indirectly the ultraproduct construction.[1] However, the use of a non-standard model from the beginning simplifies things considerably. This is analogous to the situation in non-standard analysis: while proofs using infinitesimals (and other features of non-standard analysis) can be translated into the $\epsilon - \delta$ formalization, the intuition or clarity of the original argument may be lost in the translation.

Random variables induce probabilistic distributions and probabilities of events. In particular, two random variables may be neither equal nor unequal; rather they may be equal with some probability. However, there is a fundamental problem: probabilities cannot be used as truth-values if classical logic is to be preserved. The (almost) right choice for the truth-value is the subset of the sample space consisting of those samples for which the two random variables are equal. At the heart of our construction is the realization that we can employ a bit of non-standard analysis (namely Loeb's measure: Loeb [80]) at this point: if one identifies two such truth-values (subsets of the sample space) if their symmetric difference has an infinitesimal measure one gets a *complete* Boolean algebra – this is the single most important feature of our method. Evaluation of first-order formulas in complete Boolean algebras is very natural and faithful (as it is well-known from Boole [10] for propositional logic and from Rasiowa and Sikorski [90] for predicate logic).

[1] This construction is explained in a way accessible to readers without a basic logic education in the Appendix.

The models we get are not classical but are Boolean-valued.[2] But that is perfectly sufficient for the purpose of independence results (and lengths-of-proofs lower bounds): in order to demonstrate that a sentence is not provable from a set of axioms it is enough to show that its truth value in some model is smaller than the truth value of any finite conjunction of the axioms.

Although some of the models appear interesting in their own right we interpret the construction primarily as

> *A method that reduces an independence result or a lengths-of-proofs lower bound to a combinatorial/complexity-theoretic statement about random variables.*

The combinatorial/complexity-theoretic statement we refer to here expresses that the truth-value of a particular sentence (in a particular model) is some particular value, typically $1_{\mathcal{B}}$ or $0_{\mathcal{B}}$. The validity of such a statement is a property of the particular family of random variables forming the model. For the families we consider it can often be formulated as a statement that an algorithm of a certain type can (or cannot) perform some computational task successfully for a high fraction of inputs.

Organization of the book

The book is divided into eight parts and an appendix. Part I (*Basics*) describes the general framework of the construction and develops a few basic properties of the method. This includes witnessing of quantifiers in the structures and linking the validity in the structures with the probability in the standard model. Part II (*Second-order structures*) extends the set-up to two sorted structures, with one sort for numbers and the other for bounded sets.

In Part III (*AC^0 world*) we construct two structures. The first one is a structure based on random variables computed by shallow decision trees. This is quite a rudimentary example and its basic properties are mirrored in several later models. The second structure is based on deep decision trees and it is a model of theory V_1^0. In Part IV (*$AC^0(2)$ world*) we construct an algebraic structure based on random variables defined by algebraic decision trees. This structure is a model of the theory $Q_2V_1^0$, extending V_1^0 by a bounded quantifier

[2] One can get classical models by applying a bit of logic: First apply the Löwenheim–Skolem theorem to the whole model-theoretic situation (i.e. not only the model but also the Boolean algebra and the truth valuation) to replace it by a countable one, and then apply the Rasiowa–Sikorski theorem [91] to collapse the Boolean algebra to the two-element Boolean algebra while preserving joins and meets used for defining the truth values, and hence collapsing the model to a classical one.

allowing us to count the parity of a bounded set. The key step in analyzing both the deep tree model and the algebraic model is bounded quantifier elimination. The combinatorial heart of these elimination procedures is provided by the Hastad's switching lemma and by the Razborov–Smolensky's approximation method respectively. In both Parts III and IV we use the models to derive anew a few known undefinability results, witnessing theorems and independence results for the theories. The purpose of including this material (as well as examples in Part VII) is to demonstrate that the method is a viable alternative to the usual proof-theoretic approach based on some form of a normalization of proofs.

Part V (*Towards proof complexity*) describes a general approach using the models for lengths-of-proofs lower bounds. This follows to a large extent Ajtai's method in [2], but with some important twists. In Part VI (*Proof complexity of F_d and $F_d(\oplus)$*) we use this approach to give a new proof of an exponential lower bound for PHP (the pigeonhole principle) proofs in constant-depth Frege systems. Then we discuss a long-standing open problem to prove the same lower bound also for constant-depth Frege systems with the parity gate. We do not manage to construct a model that would prove the elusive lower bound but we review some possibly relevant material about algebraic proof systems and, more importantly, we discuss in detail the issues that any construction of the desired structure has to tackle (in particular, the necessity of partially defined random variables). The models considered in this part are quite analogous to models in Parts III and IV.

The structures in Parts III–VI are second-order. In Part VII (*Polynomial-time and higher worlds*) we return to the first-order formalization and construct several models for theories like PV (polynomially verifiable), S^1_2 and T^1_2 and derive in this way some of the most important known witnessing theorems and conditional independence results in bounded arithmetic. In this part we also note a link between pseudorandom sets and a Löwenheim–Skolem phenomenon. Further, we define a model of PV naturally interpreting structural complexity results about random oracle.

In Part VIII (*Proof complexity of EF and beyond*) we first overview aims of proof complexity of strong proof systems and recall, in particular, background facts relevant to the Extended Frege proof system EF and to bounded arithmetic theories related to EF. We then expose in some detail the emerging theory of proof complexity generators aimed at constructing examples of hard tautologies. We also discuss several conjectures regarding these generators: on the hardness of proving circuit lower bounds, Razborov's conjecture about the Nisan–Wigderson generator and Extended Frege system, a conjecture about using random sparse Nisan–Wigderson generators as gadgets in gadget generators, and related Rudich's demi-bit conjecture. Finally we construct a model relevant to some of these conjectures.

The main text is supplemented by an appendix in which we present in a self-contained and quite elementary way the construction of an ultrapower extension of the standard model of natural numbers. We also try to convey, using several examples, some mental picture about the model so that even a reader who is not familiar with non-standard methods can develop some intuition and follow the arguments in the main text.

Remarks on the literature

A form of forcing has been applied in bounded arithmetic earlier. Paris and Wilkie [86] and later Ajtai [2, 3, 4] and Riis [98] used a simple variant of Robinson's model-theoretic forcing (although combined with an involved combinatorial reduction in Ajtai's [2, 3, 4]). Wilkie[3] described a construction of Boolean-valued models of the theory S^1_2 and reproved using it a relation – known previously from Cook [28] and Buss [11] – between S^1_2 and the Extended Frege proof system EF. His construction has been further extended by Krajíček [55, 57, 60, 56] to a wider context. With a slight simplification one can describe the Boolean algebras involved in these constructions as Lindenbaum algebras but not based on provable equivalence of formulas (or circuits) as they are defined classically but rather on *feasibly provable* (i.e. with proofs of polynomially bounded length) equivalence. This works well in the sense that any valid lower bound can be proved, in principle, by such a forcing. But on the other hand the algebras are defined using the notion of a feasible proof about which we are supposed to say something by the construction in the first place, and so it is in a sense a vicious circle. Takeuti and Yasumoto [103, 104] changed the feasibly provable equivalence to simply 'true equivalence' – breaking this vicious circle – but it apparently did not help much as we know very little about the power of Boolean circuits of feasible size. Most importantly, the algebras used in all these constructions are not complete but are closed only under some definable unions. That makes it very hard to use them.

Background

This is an investigation in bounded arithmetic and in proof complexity and we expect that, ideally, the reader is familiar with established basic definitions,

[3] Unpublished lecture at the International Congress on Logic, Methodology and Philosophy of Science in Moscow, 1987.

facts, methods and aims of the field. The relevant background in bounded arithmetic and proof complexity can be found in Krajíček [56] but some reader may find useful shorter explanations of some basic points in Krajíček [53, 57, 61, 62] or in an excellent survey by Pudlák [89]. Nevertheless we always briefly review the relevant theories and propositional proof systems before they are studied, and thus a reader with at least a minimal logic background should be able to study the book. In addition Chapter 27 gives some very general proof complexity background. Recently Cook and Nguyen [30] offered an excellent exposition of basic theories and their relations to proof systems. Propositional proof systems and their complexity are also treated by Clote and Kranakis [25]. A reader looking for a background in model theory, and non-standard models in particular, may consult Chang and Keisler [21], Marker [83] (ultrapowers are there in Exercises 2.5.19 – 2.5.22 and 4.5.37) or Kaye [52].

Despite the natural character and simplicity of Boolean-valued models they were discovered only in the late 1960s by P. Vopěnka, and by D. Scott and R. Solovay as their versions of Cohen's forcing; the paper by Scott [100] is a beautiful exposition of the basic ideas aimed at non-logicians, the best to date (it also contains detailed bibliographical/historical comments[4]). The paper by Scott [100] as well as virtually all later expositions (e.g. in Takeuti and Zaring [105] or in Jech [49]) consider only Boolean-valued models of set theory. Mansfield [82] attempts a general theory but concentrates on model-theoretic properties of the class of such models, as opposed to properties of particular models, and gives a version that yields only elementary extensions and hence is unsuitable for independence results.

[4] Takeuti reports in [103] that Gödel recognized in Boolean-valued models a model-theoretic version of a reinterpretation of logical operations that he had developed earlier but had never used for independence results as it was too complicated.

PART I

Basics

1
The definition of the models

1.1 The ambient model of arithmetic

Let L_{all} be the language containing symbols for every relation and function on the natural numbers $\mathbf{N}$; each symbol from L_{all} has a canonical interpretation in $\mathbf{N}$. Let $\mathcal{M}$ be an $\aleph_1$-saturated model[1] of the true arithmetic in the language L_{all}. Such a model exists by general model-theoretic constructions; see Hodges [43]. Definable sets mean definable with parameters, unless specified otherwise.

The $\aleph_1$-saturation implies the following:

(1) If a_k, $k \in \mathbf{N}$, is a countable family of elements of $\mathcal{M}$ then there exists a non-standard $t \in \mathcal{M}$ and a sequence $(b_i)_{i<t} \in \mathcal{M}$ such that $b_k = a_k$ for all $k \in \mathbf{N}$.
We shall often denote this sequence of length t simply $(a_i)_{i<t}$.

For example, if all elements $\{a_k\}_{k \in \mathbf{N}}$ obey some definable property P then – by induction in $\mathcal{M}$ (aka overspill, see the Appendix) – also some b_s with a non-standard index $s < t$ will obey P. Such an element b_s will serve well as 'a limit' (interpreted here informally) of the sequence $\{a_k\}_{k \in \mathbf{N}}$.

Another property implied by the $\aleph_1$-saturation (and equal to it if we used a countable language) is the following:

(2) If A_k, $k \in \mathbf{N}$, is a countable family of definable subsets of $\mathcal{M}$ such that $\bigcap_{i<k} A_i \neq \emptyset$ for all $k \geq 1$, then $\bigcap_k A_k \neq \emptyset$.
However, the intersection $\bigcap_k A_k$ does not need to be definable in $\mathcal{M}$.

These two statements are essentially the only consequences of the $\aleph_1$-saturation that we will use.

[1] In the Appendix we give an elementary and self-contained construction (the so-called ultrapower) of such a model.

The ambient model $\mathcal{M}$ will suffice for our purposes everywhere in this book. However, in general we could take for $\mathcal{M}$ an $\aleph_1$-saturated elementary extension of $\mathbf{N}$ in a many sorted language having names not only for all elements of $\mathbf{N}$ and relations and functions on $\mathbf{N}$ as L_{all} has, but also names for all families of sets, families of families of sets, etc., for the whole so-called *superstructure* (this is commonly done in non-standard analysis and this terminology is used there). In such a rich model the properties above would hold also for sequences of sets, families, etc.

1.2 The Boolean algebras

Let $\Omega \in \mathcal{M}$ be an arbitrary infinite set called a **sample space**. As it is an element of $\mathcal{M}$, it is $\mathcal{M}$-finite. Let $N = |\Omega|$ be the size of Ω in the sense of $\mathcal{M}$. It is necessarily non-standard.

Let $\mathcal{A} := \{A \in \mathcal{M} \mid A \subseteq \Omega\}$. This is a Boolean algebra but not a σ-algebra as the class of definable sets is not closed under all countable unions (for example, while it contains all singletons it does not contain the countable set of those elements of Ω having only standardly many predecessors in Ω). The **counting measure** (i.e. the uniform probability) on $\mathcal{A}$ is defined by:

$$A \in \mathcal{A} \rightarrow \ |A|/N.$$

Its values are the $\mathcal{M}$-rationals. A positive $\mathcal{M}$-rational is called **infinitesimal** if it is smaller that all fractions $\frac{1}{k}$, $k \in \mathbf{N}$.

Define an ideal $\mathcal{I} \subseteq \mathcal{A}$ by:

$$A \in \mathcal{I} \ \text{ iff } \ |A|/N \ \text{ is infinitesimal.}$$

$\mathcal{I}$ is not definable in $\mathcal{M}$ (otherwise the set of natural numbers $\mathbf{N}$ would be definable, violating the overspill in $\mathcal{M}$). Using $\mathcal{I}$ define a Boolean algebra $\mathcal{B} := \mathcal{A}/\mathcal{I}$.

The induced measure on $\mathcal{B}$ (the so-called Loeb's measure) will be denoted μ. Hence $\mu(b)$ for $b \in \mathcal{B}$ is the standard part of $|B|/N$ (i.e. the unique standard real infinitesimally close to it) for any $B \in \mathcal{A}$ such that $B/\mathcal{I} = b$. It is a measure in the ordinary sense: The values of μ lie in the reals $\mathbf{R}$. It is σ-additive and a strict measure: $\mu(b) > 0$ if $b \neq 0_{\mathcal{B}}$.

The following key lemma is a combination of two well-known facts, one from non-standard analysis and one from measure theory.

Lemma 1.2.1 *$\mathcal{B}$ is a complete Boolean algebra.*

Proof: First, as in the construction of Loeb's measure Loeb [80], we use the $\aleph_1$-saturation to show that

Claim 1: *$\mathcal{B}$ is a σ-algebra and the measure μ is σ-additive.*

To establish the claim let $\{b_k\}_{k\in\mathbf{N}}$ be a countable subset of $\mathcal{B}$. Assume $b_k = B_k/\mathcal{I}$ for B_ks from $\mathcal{A}$. We may assume, without loss of generality, that $B_0 \subseteq B_1 \subseteq \ldots$. It is enough to find $C \in \mathcal{A}$ such that $B_k \subseteq C$ for all $k \in \mathbf{N}$ and $\mu(C) = \lim_{k\to\infty} \mu(B_k)$. It holds that for any $k \geq 1$ there is $n_k \geq 1$ such that for all $m > \ell \geq n_k$

$$\frac{|B_\ell|}{N} \leq \frac{|B_m|}{N} \leq \frac{|B_\ell|}{N} + \frac{1}{k}.$$

We may assume, by taking subsequence $\{B_{n_k}\}_{k\in\mathbf{N}}$, that $n_k = k$.

Take a non-standard extension $\{B_i\}_{i<t}$ of $\{B_k\}_{k\in\mathbf{N}}$ guaranteed to exist by $\aleph_1$-saturation (each B_k is an element of $\mathcal{M}$). Consider the following property P parameterized by s:

$$B_s \in \mathcal{A} \wedge \forall i \leq s; B_i \subseteq B_s \wedge \frac{|B_i|}{N} \leq \frac{|B_s|}{N} \leq \frac{|B_i|}{N} + \frac{1}{i}.$$

Property P is obeyed by all standard s and hence, by induction in $\mathcal{M}$, also by some non-standard $s_0 < t$.

It is easy to verify that $C := B_{s_0}$ has the required properties.

Claim 2: *$\mathcal{B}$ satisfies the ccc condition: any antichain is at most countable.*

This holds because the measure is strictly positive, i.e. $\mu(b) > 0$ for $b \neq 0_{\mathcal{B}}$: Any antichain can contain only finitely many (non-zero) elements with measure in each interval $(1/(n+1), 1/n]$, $n \geq 1$.

As a consequence we get

Claim 3: *Any family of elements of $\mathcal{B}$ has the same set of upper bounds as one of its countable subfamilies.*

To see this, note that a family has the same set of upper bounds as the ideal it generates which in turn has the same set of upper bounds as any maximal antichain it contains – such antichains are countable by ccc (Claim 2), and each element of the antichain is majorized by a union of a finite number of elements of the family).

We can conclude the proof that $\mathcal{B}$ is complete: the union of any family can be defined as the union of a countable subfamily (Claim 3), and any countable family has a union by σ-additivity (Claim 1). □

We conclude this section by pointing out two facts that are generally *not true* in $\mathcal{B}$. The first observation is that the quotient operation $\ldots/\mathcal{I}$ that defines $\mathcal{B}$ from $\mathcal{A}$ does not commute with infinite unions: it is not true that it would generally hold that $(\bigcup_i A_i)/\mathcal{I} = \bigvee_i (A_i/\mathcal{I})$. In fact, the left-hand side may not be defined (i.e. the union may not be in $\mathcal{A}$), and, even when it is, the equality may not hold (e.g. if A_is run over all singleton subsets of Ω then the left-hand side is $\Omega/\mathcal{I} = 1_{\mathcal{B}}$ while the right-hand is $\bigvee_i 0_{\mathcal{B}} = 0_{\mathcal{B}}$).

The second failure is well-known: the infinite distributive law,

$$\bigwedge_{i\in I}\bigvee_{j\in J} b_{ij} = \bigvee_{f}\bigwedge_{i\in I} b_{i,f(i)},$$

where f range over all functions from I to J, may not hold (although the left-hand side always majorizes the right-hand side). However, it holds when I is finite.

Furthermore, for measure algebras (and hence for $\mathcal{B}$, see Chapter 2) the *weak* distributive law,

$$\bigwedge_{i\in I}\bigvee_{j\in J} b_{ij} = \bigvee_{g}\bigwedge_{i\in I}\bigvee_{j\in g(i)} b_{i,j},$$

holds, where g is a map assigning to elements of I finite subsets of J.

See Takeuti and Zaring [105] or Jech [49] for details.

1.3 The models $K(F)$

Let Ω, $\mathcal{A}$, $\mathcal{I}$ and $\mathcal{B}$ be as in the previous section.

Let $L \subseteq L_{\text{all}}$ and let F be a non-empty set of functions such that:

1. $F \subseteq \mathcal{M}$.
2. The domain of any function $\alpha \in F$ is Ω (hence $\alpha : \Omega \to \mathcal{M}$).
3. F is closed under L-functions and contains all L-constants.

The L-functions are interpreted by composition:

$$f(\alpha_1,\ldots,\alpha_k)(\omega) := f(\alpha_1(\omega),\ldots,\alpha_k(\omega)),$$

for an L-function f of arity $k \geq 1$.

Any such F will be called an **L-closed family** of random variables. Note, in particular, that F itself need not be $\mathcal{M}$-definable; in fact, in almost all our applications it will not be.

Definition 1.3.1 Assume that F is an L-closed family of random variables. $K(F)$ will denote a Boolean-valued L-structure defined as follows.

The universe of $K(F)$ is F. The Boolean valuation of L-sentences with parameters from F has values in $\mathcal{B}$ and is given by the following inductive conditions:

- $[\![\alpha = \beta]\!] := \{\omega \in \Omega \mid \alpha(\omega) = \beta(\omega)\}/\mathcal{I}$.
- $[\![R(\alpha_1,\ldots,\alpha_k)]\!] := \{\omega \in \Omega \mid R(\alpha_1(\omega),\ldots,\alpha_k(\omega))\}/\mathcal{I}$, for any L-relation R.
- $[\![\ldots]\!]$ commutes with $\neg, \vee, \wedge$.
- $[\![\exists x A(x)]\!] := \bigvee_{\alpha \in F} [\![A(\alpha)]\!]$.
- $[\![\forall x A(x)]\!] := \bigwedge_{\alpha \in F} [\![A(\alpha)]\!]$.

Note that the values of $[\![\exists x A(x)]\!]$ and $[\![\forall x A(x)]\!]$ are well-defined as $\mathcal{B}$ is complete by Lemma 1.2.1. The notation $K(F)$ is sound as Ω is determined by F and $\mathcal{A}$, $\mathcal{I}$ and $\mathcal{B}$ are determined by Ω.

1.4 Valid sentences

We say that a sentence A is **valid** in $K(F)$ iff (if and only if) its Boolean value is $1_\mathcal{B}$. It is straightforward to verify that all axioms of first-order predicate calculus in any particular formalization, including axioms of equality, are valid in any $K(F)$, and that the value $1_\mathcal{B}$ is preserved by inference rules. Further, an implication $B \to A$ is valid iff the truth value of the succedent majorizes the truth value of the antecedent:

$$[\![B]\!] \leq [\![A]\!].$$

Hence we have the following.

Lemma 1.4.1 *Let T be a set of L-sentences and A an L-sentence (parameters from $K(F)$ are allowed). If A is provable from T in predicate calculus with equality then*

$$\bigwedge_{B \in T_0} [\![B]\!] \leq [\![A]\!]$$

for some finite $T_0 \subseteq T$.

In particular, if A is logically valid then $[\![A]\!] = 1_\mathcal{B}$ (as $\bigwedge \emptyset = 1_\mathcal{B}$), i.e. A is valid in $K(F)$.

Although we are primarily interested in models of bounded arithmetic theories, and functions in the families F will typically have a small computational complexity, note that $K(F)$ can be a model of a very strong theory. For example, taking for F all functions (in $\mathcal{M}$) with domain Ω yields a model where $\mathrm{Th}(\mathbf{N})$ is valid.

For Γ a prefix class of formulas and $L \subseteq L_{\mathrm{all}}$ let $\mathrm{Th}_{\Gamma}(L)$ denote the set of true Γ-sentences valid in the canonical L-structure on $\mathbf{N}$. The following lemma is obvious.

Lemma 1.4.2 *Let F be an L-closed family. Then* $\mathrm{Th}_{\forall}(L)$ *is $K(F)$-valid.*
If F contains (constant functions) $\mathbf{N}$ *then* $\mathrm{Th}_{\exists\forall}(L)$ *is $K(F)$-valid.*

This seems to block a priori constructions of non-Π^b_1-elementary extensions which are needed for lengths-of-proofs lower bounds. However, there is a simple way out: a statement that can be expressed by a Π^b_1-formula or even an atomic formula in one language does not need to be so expressible in another language. For example, if we have a function symbol for an algorithm checking that a truth assignment satisfies a set of clauses then such a statement is expressible by an atomic formula. But without such a symbol we may need a Π^b_2-formula to express it (every clause contains a satisfied literal).

Lemma 1.4.2 will also be used in Chapter 4 in a crucial way to foster a link between the truth in $\mathbf{N}$ and the validity in $K(F)$.

1.5 Possible generalizations

We could have started the construction of $\mathcal{B}$ from any probability space $(\Omega, \mathcal{A}, \nu)$ that is definable in $\mathcal{M}$. In particular, both sets Ω and $\mathcal{A}$ have to be definable in $\mathcal{M}$ (the definability of the whole $\mathcal{A}$ – the definability of each element of $\mathcal{A}$ is not enough – is needed for Loeb's construction) and ν must be a definable finitely additive measure on $\mathcal{A}$ (with values in $\mathcal{M}$-rationals). One also needs to add a condition on αs from F: each $\alpha^{(-1)}(a)$ for $a \in \mathcal{M}$ has to be measurable, i.e. in $\mathcal{A}$.

An example of this more general space is a probability space on an $\mathcal{M}$-finite Ω with a non-uniform distribution. This will be used, for example, in Chapter 25.

An example of a space with an $\mathcal{M}$-infinite Ω is $\Omega := \{w \in \mathcal{M} \mid |w| \geq n\}$, $\mathcal{A}$ consisting of all $\mathcal{M}$-finite Boolean combinations of r.e. (recursively enumerable) subsets of Ω (defined using a code for the Boolean combination and a universal Σ^0_1-formula) and measure ν giving to a string w the weight $2^{n-1-2|w|}$.

However, we have no application for these more general constructions here and we restrict to the counting measure on $\mathcal{M}$-finite Ω.

Two random variables α, β on $\{0,1\}^n$ are 'almost equal' (i.e. the formula $\alpha = \beta$ is $K(F)$-valid) if they differ for an infinitesimal fraction of samples $\omega \in \{0,1\}^n$. In complexity theory, however, a stronger condition would be used: the random variables should differ in a negligible fraction, i.e. less than n^{-k}, for any $k \in \mathbf{N}$. Whenever necessary (not in this book, however) this can be remedied analogously to complexity theory: build the model from the same random variables but computed on a tuple of independent samples.

More precisely, let F be an L-closed family on Ω, and let $t \geq 1$, $t \in \mathcal{M}$, be arbitrary. For $\alpha \in F$ define $\alpha^{(t)} : \Omega^t \to M^t$ by:

$$\alpha^{(t)}(\omega_1, \ldots, \omega_t) := (\alpha(\omega_1), \ldots, \alpha(\omega_t)).$$

Denote by $F^{(t)}$ the collection of all $\alpha^{(t)}$; it is also an L-closed family.

We interpret an L-relation R on elements $a^i = (a^i_1, \ldots, a^i_t) \in \mathcal{M}^t$, $i \leq k$, as

$$R(a^1, \ldots, a^k) := \bigwedge_{j \leq t} R(a^1_j, \ldots, a^k_j),$$

i.e. we assign the Boolean value by the equation:

$$[\![R((\alpha^1)^{(t)}, \ldots, (\alpha^k)^{(t)})]\!] := \{(\omega_1, \ldots, \omega_t) \in \Omega^t \mid \bigwedge_{j \leq t} R(\alpha^1(\omega_j), \ldots, \alpha^k(\omega_j))\}/\mathcal{I}.$$

Let $\mathcal{A}^t$ be the t-product of $\mathcal{A}$, a probability space on Ω^t, and let $\mathcal{B}^{(t)}$ be the complete Boolean algebra resulting from the construction in Section 1.2. (It is not the t-folded product of $\mathcal{B}$; in fact, such $\mathcal{B}^t$ is not even defined when t is non-standard.) The $F^{(t)}$-formulas are evaluated in $\mathcal{B}^{(t)}$.

Then for $\alpha^{(t)} = \beta^{(t)}$ to be valid in $K(F^{(t)})$ it is not sufficient that $p := \mathrm{Prob}_{\omega \in \Omega}[\alpha(\omega) \neq \beta(\omega)]$ is infinitesimal; it must be that $p \cdot t$ is infinitesimal.

2
Measure on $\mathcal{B}$

Boolean algebra $\mathcal{B}$ carries a strict probabilistic measure μ. We shall interpret it in this chapter as a metric and also link it with probability.

2.1 A metric on $\mathcal{B}$

The map

$$A \longrightarrow \mu([\![A]\!])$$

assigns real numbers to L-sentences, and satisfies

- $0 \leq \mu([\![A]\!]) \leq 1$
- $\mu([\![\neg A]\!]) = 1 - \mu([\![A]\!])$
- $\mu([\![A \vee B]\!]) + \mu([\![A \wedge B]\!]) = \mu([\![A]\!]) + \mu([\![B]\!])$

and, with suitable F (see Chapter 3), also

- $\mu([\![\exists x A(x)]\!]) = \sup_{\alpha \in F} \mu([\![A(\alpha)]\!])$ and
- $\mu([\![\forall x A(x)]\!]) = \inf_{\alpha \in F} \mu([\![A(\alpha)]\!]))$.

Further it satisfies the all-important inequality

- $\mu([\![A]\!]) \leq \mu([\![B]\!])$ if A logically implies B.

However, the number $\mu([\![A]\!])$ cannot be interpreted as the probability of A. The passage from the Boolean value to probability is more subtle (Section 2.2). The appropriate interpretation of this number seems to be in terms of a metric on $\mathcal{B}$. Define for $a, b \in \mathcal{B}$:

$$d(a,b) := \mu(a \Delta b)$$

where $a \Delta b$ is the symmetric difference $(a \wedge \neg b) \vee (\neg a \wedge b)$, denoted in Boolean complexity also as $a \oplus b$ (the sum).

The following lemma is verified by looking at the Venn diagram.

Lemma 2.1.1 *Function d is a metric on $\mathcal{B}$. Further, the following equality and two inequalities hold for any $a, a', b, b' \in \mathcal{B}$:*

1. $d(a,b) = d(\neg a, \neg b)$.
2. $d(a \wedge b, a' \wedge b') \leq d(a,a') + d(b,b')$.
3. $d(a \vee b, a' \vee b') \leq d(a,a') + d(b,b')$.

2.2 From Boolean value to probability

In this section we investigate a relation between the probability of a sentence in $\mathcal{M}$ and the μ-measure of its Boolean value in $K(F)$.

The following lemma follows from Definition 1.3.1 and from the fact that the quotient operation $A \in \mathcal{A} \to A/\mathcal{I} \in \mathcal{B}$ commutes with $\neg, \vee, \wedge$ (binary).

Lemma 2.2.1 *Let F be an L-closed family, let $A(x)$ be an open L-formula with the only free variable x, and let $\alpha \in F$.*

Then for all standard $\epsilon > 0$

$$\operatorname*{Prob}_{\omega \in \Omega}[A(\alpha(\omega))] \geq \mu([\![A(\alpha)]\!]) - \epsilon.$$

(The probability expression is defined in $\mathcal{M}$.)

In order to be able to apply Lemma 3.3.3 later we need some extension of Lemma 2.2.1 to universal formulas. However, in general it is not true that $\mathrm{Prob}_{\omega \in \Omega}[\forall z A(\alpha(\omega), z)]$ can be bounded from below by a term in $\mu([\![\forall z A(\alpha, z)]\!])$. For example, it can be that $\forall x \exists! z \neg A(x,z)$ but that no $\gamma \in F$ finds such z for more than a negligible fraction of xs. Then $[\![\forall z A(\mathrm{id}_\Omega, z)]\!] = 1_{\mathcal{B}}$ while $\mathrm{Prob}_{\omega \in \Omega}[\forall z A(\omega, z)] = 0$.

The way out is to choose suitable families F. Call any function $\xi \in L_{\mathrm{all}}$ satisfying:

$$\mathbf{N} \models \forall x\, [\exists z \neg A(x,z) \to \neg A(x, \xi(x))]$$

a counter-example function for $\forall z A(x,z)$.

Lemma 2.2.2 *Assume that $A(x,z)$ is an open L-formula, ξ is a counter-example function for $\forall z A(x,z)$, and F is an L-closed family closed under ξ.*

Then for any $\alpha \in F$ and any standard $\epsilon > 0$

$$\operatorname*{Prob}_{\omega \in \Omega}[\forall z A(\alpha(\omega), z)] \geq \mu([\![\forall z A(\alpha, z)]\!]) - \epsilon.$$

Proof: Let $[\![\forall z A(\alpha, z)]\!] = b$. For all $\gamma \in F$ we have $[\![A(\alpha, \gamma)]\!] \geq b$, so

$$\operatorname*{Prob}_{\omega \in \Omega}[A(\alpha(\omega), \gamma(\omega))] \geq \mu(b) - \epsilon,$$

for any $\epsilon > 0$ by Lemma 2.2.1.

For a particular $\gamma(\omega) := \xi(\alpha(\omega))$ we also have

$$(\forall z A(\alpha(\omega), z)) \equiv A(\alpha(\omega), \gamma(\omega)),$$

so

$$\operatorname*{Prob}_{\omega \in \Omega}[\forall z A(\alpha(\omega), z)] = \operatorname*{Prob}_{\omega \in \Omega}[A(\alpha(\omega), \gamma(\omega))]$$

and the lemma follows. □

3
Witnessing quantifiers

Consider an L-sentence A of the form

$$\exists x_1 \forall y_1 \dots \exists x_k \forall y_k B(x_1, y_1, \dots, x_k, y_k)$$

with B open. Let K be an L-structure. Two players, $\exists$-player and $\forall$-player, pick sequentially witnesses $\alpha_1, \beta_1, \dots, \alpha_k, \beta_k \in F$ to the quantifiers. The $\exists$-player wants to make the truth value $[\![B(\alpha_1, \beta_1, \dots, \alpha_k, \beta_k)]\!]$ as large as possible by selecting $\alpha_1, \dots, \alpha_k$ suitably, while the $\forall$-player's objective is the opposite, to make it as small as possible by picking $\beta_1, \dots, \beta_k$ suitably.

If K were a classical structure (i.e. if $\mathcal{B}$ were the 2-element algebra) the situation would be simple: if A is true then the $\exists$-player has a strategy to achieve value $1_{\mathcal{B}}$, and if A is false the $\forall$-player has a strategy to reach value $0_{\mathcal{B}}$. However, if K is a Boolean-valued structure there may not be optimal moves $\alpha_1, \beta_1, \dots$ that would ensure that

$$[\![B(\alpha_1, \beta_1, \dots, \alpha_k, \beta_k)]\!] = [\![\exists x_1 \forall y_1 \dots \exists x_k \forall y_k B(x_1, y_1, \dots, x_k, y_k)]\!]$$

and, in general, the players need to collaborate to get as close (in terms of the metric on $\mathcal{B}$) to

$$[\![\exists x_1 \forall y_1 \dots \exists x_k \forall y_k B(x_1, y_1, \dots, x_k, y_k)]\!]$$

as possible.

In this chapter we will investigate under which conditions posed on F we can guarantee the existence of suitable moves (often called 'witnesses') $\alpha_1, \beta_1, \dots, \alpha_k, \beta_k$. Let us mention two reasons why we are interested in this.

Consider a sentence A of the form $\forall x \exists y B(x, y)$, with B open. Assume that $\mathrm{id}_\Omega \in F$, where $\Omega = \{0,1\}^n$. If $[\![A]\!] = b$ then $[\![\exists y B(\mathrm{id}_\Omega, y)]\!] \geq b$. Assume that we can witness y by some $\alpha \in F$ such that

$$[\![\exists y B(\mathrm{id}_\Omega, y)]\!] = [\![B(\mathrm{id}_\Omega, \alpha)]\!].$$

Then, by Lemma 2.2.1

$$\mathrm{Prob}_{\omega \in \Omega}[B(\omega, \alpha(\omega)] \geq \mu(b) - \epsilon,$$

for all standard $\epsilon > 0$. Hence we have reduced the question of approximating $[\![A]\!]$ (in $\mathcal{B}$ as a metric space) to a question about the witnessing power of F-functions in the ambient model $\mathcal{M}$, and consequently also in $\mathbf{N}$.

For the second example assume that we can find witnesses $\alpha_1, \ldots, \beta_k$ to quantifiers in $A(\gamma) := \exists x_1 \forall y_1 \ldots \exists x_k \forall y_k B(x_1, y_1, \ldots, x_k, y_k, \gamma)$ such that

$$[\![B(\alpha_1, \beta_1, \ldots, \alpha_k, \beta_k, \gamma)]\!] = [\![\exists x_1 \forall y_1 \ldots \exists x_k \forall y_k B(x_1, y_1, \ldots, x_k, y_k, \gamma)]\!]$$

(B is again open and has a parameter $\gamma \in F$). The left-hand side value can be computed explicitly:

$$\{\omega \mid B(\alpha_1(\omega), \beta_1(\omega), \ldots, \alpha_k(\omega), \beta_k(\omega), \gamma(\omega))\}/\mathcal{I}.$$

So in this case we are able to represent the truth value of an arbitrarily complex formula by a set defined by an open formula. Such computation of the truth values of a complex formula by the truth value of a simpler one is, in principle, possible because the value $[\![A(\gamma)]\!]$ has, in general, nothing to do with

$$\{\omega \mid \exists x_1 \forall y_1 \ldots \exists x_k \forall y_k B(x_1, y_1, \ldots, x_k, y_k, \gamma(\omega))\}/\mathcal{I}.$$

When we shall in future quantify in statements over $\epsilon > 0$ or $\delta > 0$ it is tacitly over *standard* ϵ or δ. Infinitesimal positive quantity will always be written as $1/t$ with t non-standard.

3.1 Propositional approximation of truth values

Let A be a sentence (possibly with parameters) of the form $\exists x B(x)$; B does not need to be open here. The next statement was used already in the proof of Lemma 1.2.1 and we state it here for completeness. It is a consequence of the ccc property of $\mathcal{B}$ (the second part is a consequence of the σ-additivity of μ).

Lemma 3.1.1 *There are countably many $\alpha_k \in F$, $k \in \mathbf{N}$, such that*

$$[\![\exists x B(x)]\!] = \bigvee_{k \in \mathbf{N}} [\![B(\alpha_k)]\!].$$

In particular, for any $\epsilon > 0$ there are finitely many $\alpha_1, \ldots, \alpha_\ell \in F$ such that

$$d\left([\![\exists x B(x)]\!], \bigvee_{k \leq \ell} [\![B(\alpha_k)]\!]\right) < \epsilon.$$

An analogous statement holds also for approximating the truth value of a formula starting with $\forall$ by $\bigwedge$ of its instances.

Iterating Lemma 3.1.1 yields the following statement.

Lemma 3.1.2 *Let A be an L-sentence of the form*

$$\exists x_1 \forall y_1 \ldots \exists x_k \forall y_k B(x_1, y_1, \ldots, x_k, y_k)$$

with B open (parameters are allowed). Let $\epsilon > 0$ be arbitrary.

Then there are $\ell \in \mathbf{N}$ and elements of F

$$\alpha_1^{i_1}, \beta_1^{i_1, j_1}, \alpha_2^{i_1, j_1, i_2}, \ldots, \beta_k^{i_1, \ldots, j_k}$$

with $i_1, \ldots, j_k < \ell$, such that the distance between $[\![A]\!]$ and

$$\bigvee_{i_1} \bigwedge_{j_1} \cdots \bigvee_{i_k} \bigwedge_{j_k} [\![B(\alpha_1^{i_1}, \beta_1^{i_1, j_1}, \ldots, \beta_k^{i_1, \ldots, j_k})]\!]$$

is less than ϵ.

In particular, the distance between $[\![A]\!]$ and

$$\left\{\omega \in \Omega \mid \bigvee_{i_1} \bigwedge_{j_1} \cdots \bigvee_{i_k} \bigwedge_{j_k} B(\alpha_1^{i_1}(\omega), \beta_1^{i_1, j_1}(\omega), \ldots, \beta_k^{i_1, \ldots, j_k}(\omega))\right\} / \mathcal{I}$$

is less than ϵ.

Proof: Fix small enough $\delta > 0$ (how small will be clear at the end of the proof), a standard rational. Denote:

$$C(x_1) := \forall y_1 \exists x_2 \forall y_2 \ldots B(x_1, y_1, \ldots).$$

By Lemma 3.1.1 there are $\alpha_1^0,\ldots,\alpha_1^{u-1} \in F$, $u \in \mathbf{N}$, such that

$$d\left([\![A]\!], \bigvee_{i<u} [\![C(\alpha_1^i)]\!] \right) \leq \delta.$$

Next denote for $i < u$

$$D_i(y_1) := \exists x_2 \forall y_2 \ldots B(\alpha_1^i, y_1, x_2, y_2, \ldots).$$

By Lemma 3.1.1 again there are $\beta_1^{i,j} \in F, j < v$ for some standard v, such that

$$d\left([\![C(\alpha_1^i)]\!], \bigwedge_{j<v} [\![D_i(\beta_1^{i,j})]\!] \right) \leq \delta/u$$

for all $i < u$. By Lemma 2.1.1

$$d\left([\![A]\!], \bigvee_{i<u} \bigwedge_{j<v} [\![\exists x_2 \forall y_2 \ldots B(\alpha_1^i, \beta_1^{i,j}, x_2, y_2, \ldots)]\!] \right) \leq 2\delta.$$

Proceeding analogously for $\exists x_2, \forall y_2, \ldots$ we find in F elements

$$\alpha_1^{i_1}, \beta_1^{i_1,j_1}, \alpha_2^{i_1,j_1,i_2}, \ldots, \beta_k^{i_1,\ldots,j_k}$$

with all $i_1,\ldots,j_k$ ranging over some standard finite domain such that the distance of $[\![A]\!]$ from

$$\bigvee_{i_1} \bigwedge_{j_1} \ldots [\![B(\alpha_1^{i_1}, \beta_1^{i_1,j_1}, \ldots, \beta_k^{i_1,\ldots,j_k})]\!]$$

is at most $2k\delta$. However, this truth value equals, modulo the ideal $\mathcal{I}$, the set of all $\omega \in \Omega$ satisfying (as evaluated in $\mathcal{M}$) the Boolean formula formed from formulas $B(\alpha_1^{i_1}(\omega), \beta_1^{i_1,j_1}(\omega), \ldots, \beta_k^{i_1,\ldots,j_k}(\omega))$:

$$\bigvee_{i_1} \bigwedge_{j_1} \ldots B(\alpha_1^{i_1}(\omega), \beta_1^{i_1,j_1}(\omega), \ldots, \beta_k^{i_1,\ldots,j_k}(\omega)).$$

By picking $\delta > 0$ less than $\epsilon/2k$ we get the lemma. □

3.2 Witnessing in definable families

Our next lemma gives a sample statement and a proof whose variants will be used repeatedly. The construction shows that $\aleph_1$-saturation can be used in a place where in set theory we could have used a straightforward induction on $k \in \mathbf{N}$.[1]

Lemma 3.2.1 *Assume that F is definable in $\mathcal{M}$. Let $b_k := B_k/\mathcal{I}$ for some $B_k \in \mathcal{A}$ and $\alpha_k \in F$, for all $k \in \mathbf{N}$. Assume that $b_k \wedge b_\ell = 0_\mathcal{B}$ for all $k \neq \ell \in \mathbf{N}$ (i.e. b_ks form an antichain).*

Further assume that for all $k \in \mathbf{N}$ there is $\beta \in F$ such that for all $\ell = 1, \dots, k$

$$(1) \qquad \beta(\omega) = \alpha_\ell(\omega) \text{ for } \omega \in B_\ell \setminus \bigcup_{i<\ell} B_i.$$

Then there exists $\beta \in F$ such that for all $k \in \mathbf{N}$

$$(2) \qquad [\![\beta = \alpha_k]\!] \geq b_k.$$

Proof: Let $S = \{(B_i, \alpha_i)\}_{i<t}$ be a non-standard extension of $\{(B_k, \alpha_k)\}_{k \in \mathbf{N}}$ provided by $\aleph_1$-saturation. Using S, and the fact that F is definable, we can define for all $k \in \mathbf{N}$ the set C_k of all $\beta \in F$ satisfying condition (1).

Clearly $C_0 \supseteq C_1 \supseteq \dots$ and each $C_k \neq \emptyset$ (by the hypothesis of the lemma). Hence, again by $\aleph_1$-saturation, $\bigcap_{k \in \mathbf{N}} C_k \neq \emptyset$, and any β from the intersection satisfies condition (2). □

We derive a simple corollary about witnessing of quantifiers for particular families F. First a definition:

Definition 3.2.2 Family F is **closed under definition by cases** if for any $\alpha_0, \alpha_1 \in F$ and any $B \in \mathcal{A}$ there is $\beta \in F$ such that

$$\beta(\omega) = \begin{cases} \alpha_0(\omega) & \text{if } \omega \in B \\ \alpha_1(\omega) & \text{otherwise.} \end{cases}$$

Clearly hypothesis (1) of Lemma 3.2.1 is satisfied for families closed under definitions by cases.

Lemma 3.2.3 *Assume that F is definable in $\mathcal{M}$ and that it is closed under definitions by cases.*

[1] Paradoxically there is 'no arithmetic in arithmetic', meaning that in arithmetic we do not have the set of natural numbers: PA is equivalent (in the sense of mutual interpretability) to ZFC with the Axiom of Infinity replaced by its negation. This seems to block a transplantation of Cohen's forcing to arithmetic.

Then for each sentence of the form $\exists x C(x)$ *there is an* $\alpha \in F$ *such that*

$$[\![\exists x C(x)]\!] = [\![C(\alpha)]\!]$$

(C need not be open).

In fact, for each sentence A of the form

$$\exists x_1 \forall y_1 \ldots \exists x_k \forall y_k B(x_1, y_1, \ldots, x_k, y_k)$$

(B arbitrary, possibly with parameters), there are $\alpha_1, \beta_1, \ldots, \alpha_k, \beta_k \in F$ *such that*

$$\begin{aligned}[\![A]\!] &= [\![\forall y_1 \exists x_2 \forall y_2 \ldots \exists x_k \forall y_k B(\alpha_1, y_1, \ldots, x_k, y_k)]\!] \\ &= [\![\exists x_2 \forall y_2 \ldots \exists x_k \forall y_k B(\alpha_1, \beta_1, \ldots, x_k, y_k)]\!] \\ &= \cdots = [\![\forall y_k B(\alpha_1, \beta_1, \ldots, \alpha_k, y_k)]\!] \\ &= [\![B(\alpha_1, \beta_1, \ldots, \alpha_k, \beta_k)]\!].\end{aligned}$$

Proof: The second part of the lemma is proved by induction on the number of quantifiers in A using the first part.

By Lemma 3.1.1 there are α_k, $k \in \mathbf{N}$, such that $[\![\exists x C(x)]\!] = \bigvee_{k \in \mathbf{N}} [\![C(\alpha_k)]\!]$. Define $b_0 := [\![C(\alpha_0)]\!]$ and for $k > 0$ put

$$b_k := [\![C(\alpha_k)]\!] \setminus \bigvee_{\ell < k} b_\ell.$$

By induction one can verify that for all k

$$\bigvee_{i \le k} [\![C(\alpha_i)]\!] = \bigvee_{i \le k} b_i$$

and, in particular,

$$\bigvee_{\alpha \in F} [\![C(\alpha)]\!] = \bigvee_k b_k.$$

Clearly any two different b_ks are incompatible. By Lemma 3.2.1 there is $\beta \in F$ such that

$$[\![\beta = \alpha_k]\!] \ge b_k, \quad \text{for all } k \in \mathbf{N}.$$

But then clearly

$$[\![C(\beta)]\!] \le [\![\exists x C(x)]\!] = \bigvee_{\alpha \in F} [\![C(\alpha)]\!] = \bigvee_{k \in \mathbf{N}} b_k \le [\![C(\beta)]\!].$$

The last inequality uses $b_k \leq [\![C(\alpha_k)]\!] \wedge [\![\beta = \alpha_k]\!]$ and hence $b_k \leq [\![C(\beta)]\!]$, for all $k \in \mathbf{N}$. □

Note that having witnesses $\alpha_1, \ldots, \beta_k$ such that

$$[\![A]\!] = [\![B(\alpha_1, \beta_1, \ldots, \alpha_k, \beta_k)]\!]$$

does not, in general, imply that the witnesses also obey the remaining equations stated in the lemma.

The problem with the witnessing statement 3.2.3 is that families F that we encounter never meet the two requirements:

1. F is closed under definition by cases.
2. F is definable in $\mathcal{M}$.

An example of F to keep in mind is a family F_{PV} consisting of polynomial-time functions (i.e. random variables on $\{0,1\}^n$ sampled by p-time algorithms). F_{PV} satisfies neither of the two conditions. The first condition fails for F_{PV} as a set $B \in \mathcal{A}$ from Definition 3.2.2 can be defined by a formula with quantifiers and we have no a priori polynomial upper bound on an algorithm deciding the membership in B. Family F_{PV} is also not definable in $\mathcal{M}$ as otherwise one could define $\mathbf{N}$ (as the set of k such that F_{PV} contains a function of growth faster than n^k).

We shall see in the next two sections that it is possible to relax the conditions so that Lemma 3.2.3 still holds but interesting families will satisfy the new conditions.

3.3 Definition by cases by open formulas

Definition 3.3.1 We shall say that F is **closed under definition by cases by open L-formulas** iff whenever $\alpha, \beta \in F$ and $B(x)$ is an open L-formula with free variable x then there is $\gamma \in F$ such that

$$\gamma(\omega) = \begin{cases} \alpha(\omega) & \text{if } B(\alpha(\omega)) \text{ holds} \\ \beta(\omega) & \text{otherwise.} \end{cases}$$

Note that the example family F_{PV} of polynomial-time functions is closed under definitions by cases by open L-formulas, as long as all relations and functions in L are computable by polynomial-time algorithms.

Lemma 3.3.2 *Let F be an L-closed family closed under definition by cases by open $L(F)$-formulas. Let $B(x)$ be an open L-formula (possibly with parameters from F).*

Then for every $\epsilon > 0$ there is $\alpha \in F$ such that

$$d([\![\exists x B(x)]\!], [\![B(\alpha)]\!]) < \epsilon.$$

In particular, if $\exists x B(x)$ is $K(F)$-valid then for every $\epsilon > 0$ there is $\alpha \in F$ such that

$$\mu([\![B(\alpha)]\!]) > 1 - \epsilon.$$

Proof: For any two $\beta_1, \beta_2 \in F$ there is (by the hypothesis that F is closed under definition by cases by open L-formulas) $\beta_3 \in F$ such that $\beta_3(\omega) = \beta_1(\omega)$ if $B(\beta_1(\omega))$ holds, and $\beta_3(\omega) = \beta_2(\omega)$ otherwise. Then it follows rather easily from the definition of β_3 that:

$$[\![B(\beta_3)]\!] = [\![B(\beta_1)]\!] \vee [\![B(\beta_2)]\!].$$

Combining this with Lemma 3.1.1 we get a countable sequence $\beta_0, \beta_1, \dots \in F$ such that

$$[\![B(\beta_0)]\!] \leq [\![B(\beta_1)]\!] \leq \dots$$

and such that $\lim_{k\to\infty} \mu([\![B(\beta_k)]\!]) = \mu([\![\exists x B(x)]\!])$. The lemma follows. □

Now we extend Lemma 3.3.2 to more complex formulas.

Lemma 3.3.3 *Let F be an L-closed family closed under definition by cases by open $L(F)$-formulas (L-formulas with parameters from F).*

Let $\exists x C(x)$ be an L-sentence, and let C be an arbitrary formula with parameters from F. Then for any $\epsilon > 0$ there is an $\alpha \in F$ such that

$$d([\![\exists x C(x)]\!], [\![C(\alpha)]\!]) < \epsilon.$$

In particular, if $\exists x C(x)$ is $K(F)$-valid then

$$\mu([\![C(\alpha)]\!]) > 1 - \epsilon.$$

In fact, for each sentence A of the form

$$\exists x_1 \forall y_1 \dots \exists x_k \forall y_k B(x_1, y_1, \dots, x_k, y_k)$$

(B open, possibly with parameters), there are $\alpha_1, \beta_1, \ldots, \alpha_k, \beta_k \in F$ *such that*

$$d([\![A]\!], [\![\forall y_1 \exists x_2 \forall y_2 \ldots \exists x_k \forall y_k B(\alpha_1, y_1, \ldots, x_k, y_k)]\!]) < \epsilon,$$

$$d([\![\forall y_1 \exists x_2 \ldots \exists x_k \forall y_k B(\alpha_1, y_1, x_2, \ldots, x_k, y_k)]\!],$$
$$[\![\exists x_2 \ldots \forall y_k B(\alpha_1, \beta_1, x_2, \ldots, y_k)]\!]) < \epsilon,$$

$$\ldots$$

$$d([\![\forall y_k B(\alpha_1, \beta_1, \ldots, \alpha_k, y_k)]\!], [\![B(\alpha_1, \beta_1, \ldots, \alpha_k, \beta_k)]\!]) < \epsilon.$$

Proof: The lemma appears similar to Lemma 3.2.3 where the second part followed from the first part. However, here it is the other way round: we need to prove the second part; the first part is a special case of it.

We shall describe the construction for $k = 1$, i.e. for A of the form

$$\exists x \forall y B(x, y),$$

with B open. This will display the key idea; the general case $k \geq 1$ is proved analogously by induction on the quantifier complexity of the formula.

The following claim follows from Lemma 3.3.2 applied to $\exists y \neg B(\alpha, y)$ (here we use the hypothesis that F is closed under definition by cases by open $L(F)$-formulas).

Claim 1: *For any* $\alpha \in F$ *and* $\delta > 0$ *there is* $\beta \in F$ *such that*

$$d([\![\forall y B(\alpha, y)]\!], [\![B(\alpha, \beta)]\!]) < \delta.$$

The next claim is a bit subtler.

Claim 2: *For any* $\alpha_1, \alpha_2 \in F$ *and* $\xi > 0$ *there is* $\alpha_3 \in F$ *such that*

$$d([\![\forall y B(\alpha_3, y)]\!], [\![\forall y B(\alpha_1, y)]\!] \vee [\![\forall y B(\alpha_2, y)]\!]) < \xi.$$

To prove the claim we first apply Claim 1 to α_1, α_2 and $\delta > 0$ (to be specified later) to get $\beta_1, \beta_2 \in F$ such that

$$(1) \qquad d([\![B(\alpha_i, \beta_i)]\!], [\![\forall y B(\alpha_i, y)]\!]) < \delta, \quad \text{for } i = 1, 2.$$

Define pair α_3, β_3:

$$\alpha_3(\omega), \beta_3(\omega) = \begin{cases} \alpha_1(\omega), \beta_1(\omega) & \text{if } B(\alpha_1(\omega), \beta_1(\omega)) \text{ holds} \\ \alpha_2(\omega), \beta_2(\omega) & \text{otherwise.} \end{cases}$$

Clearly $[\![B(\alpha_3,\beta_3)]\!] = [\![B(\alpha_1,\beta_1)]\!] \vee [\![B(\alpha_2,\beta_2)]\!]$. Observe that:

$$(2) \qquad d([\![\forall y B(\alpha_3,y)]\!],[\![B(\alpha_3,\beta_3)]\!]) < 3\delta.$$

This is because otherwise we could find (via Claim 1) $\gamma \in F$ such that

$$d([\![\forall y B(\alpha_3,y)]\!],[\![B(\alpha_3,\gamma)]\!]) < \delta$$

and use it to modify β_i, $i = 1$ or $i = 2$, to decrease the measure of $[\![B(\alpha_i,\beta_i)]\!]$ by more than δ, which is impossible by the choice of β_is.

Putting (1) and (2) together, and using Lemma 2.1.1, we get:

$$(3) \qquad d([\![\forall y B(\alpha_3,y)]\!],[\![\forall y B(\alpha_1,y)]\!] \vee [\![\forall y B(\alpha_2,y)]\!]) < 5\delta,$$

and taking $\delta := \xi/5$ yields the claim.

To conclude the proof of the lemma let $\alpha_1,\alpha_2,\ldots$ be a countable family such that

$$\bigvee_k [\![\forall y B(\alpha_k,y)]\!] = [\![\exists x \forall y B(x,y)]\!].$$

Put $\hat{\alpha}_1 := \alpha_1$ and apply Claim 2 to pairs $\hat{\alpha}_k,\alpha_{k+1}$ and parameters $\xi_k := \epsilon/2^{k+1}$ to get $\hat{\alpha}_{k+1}$ for $k = 1,2,\ldots$. Then

$$d\left([\![\forall y B(\hat{\alpha}_k,y)]\!], \bigvee_{j\le k}[\![\forall y B(\alpha_j,y)]\!]\right) < \epsilon/2\left(\frac{2^k-1}{2^k}\right).$$

Hence for k so large that

$$d\left(\bigvee_{j\le k}[\![\forall y B(\alpha_j,y)]\!],[\![\exists x \forall y B(x,y)]\!]\right) < \epsilon/2$$

$\hat{\alpha}_k$ is the desired element. □

3.4 Compact families

The following concept will be substituted in Section 3.5 for the definability of F used in Lemma 3.2.3.

Definition 3.4.1 Let $F \subseteq \mathcal{M}$ be a family. F is **compact** iff there exists an L_{all}-formula $H(x,y)$ such that for

$$F_a := \{b \in \mathcal{M} \mid \mathcal{M} \models H(a,b)\}$$

the following two properties hold:

1. $\bigcap_{k \in \mathbf{N}} F_k = F$.
2. $F_k \supseteq F_{k+1}$, for all $k \in \mathbf{N}$.

The second property is only technical: if $H(x,y)$ satisfies the first one then $H'(x,y) := \forall z \leq x H(z,y)$ will satisfy both (and, in fact, the second one will be satisfied for all $a \in \mathcal{M}$).

Our example family F_{PV} of polynomial-time computable functions is not compact. However, in many applications we can replace it by a bigger family F that is compact: the family of functions computable by circuits of subexponential size (here we take $\Omega = \{0,1\}^n$ and the qualification 'subexponential' means less than $2^{n^{\xi}}$, for some infinitesimal ξ, i.e. of size $2^{n^{o(1)}}$). Definable sets F_k witnessing that F is compact consist of functions computable by circuits of size less than $2^{n^{1/k}}$. Note that we switched in this example to non-uniform algorithms in order to gain the definability of F_ks. Of course, we could have produced a tighter compact envelope F' of F_{PV} by taking for F'_k, for example, the set of functions computable by circuits of size less than $n^{\log^{(k)}(n)}$ ($\log^{(k)}$ is the k-times iterated logarithm). However, the subexponential family yields stronger independence results and lower bounds.

The following lemma justifies the name **compact**: if we think of non-standard extensions of sequences as their 'limits' then it states that any sequence of elements of a compact family F has a limit in F.

Lemma 3.4.2 *Let F be a compact family. Assume C_k, $k \in \mathbf{N}$, are definable sets such that*

$$F \cap \bigcap_{\ell < k} C_\ell \neq \emptyset,$$

for all $k \geq 1$. Then

$$F \cap \bigcap_{k \in \mathbf{N}} C_k \neq \emptyset.$$

Further, if $\{\alpha_k\}_{k\in\mathbf{N}}$ *is a countable sequence of elements of* F *and* $(\alpha_i)_{i<t}$ *is its arbitrary non-standard extension then there is non-standard* $s<t$ *such that*

$$\alpha_i \in F, \quad \text{for all } i \le s.$$

Proof: Let $H(x,y)$ be the formula certifying that F is compact. Put:

$$F_a := \{b \in \mathcal{M} \mid H(a,b)\}.$$

Define sets

$$D_k := C_k \cap F_k.$$

These are definable (as F_ks are) and satisfy

$$\bigcap_{\ell<k} D_\ell \neq \emptyset$$

as

$$\bigcap_{\ell<k} D_\ell \supseteq \bigcap_{\ell<k} C_\ell \cap F \neq \emptyset.$$

Here we use $F_k \supseteq F$.

By $\aleph_1$-saturation we have:

$$\bigcap_{k\in\mathbf{N}} D_k \neq \emptyset.$$

But any $\alpha \in \bigcap_{k\in\mathbf{N}} D_k$ is, in particular, an element of $\bigcap_{k\in\mathbf{N}} F_k = F$.

For the second part and any non-standard extension $(\alpha_i)_{i<t}$ of $\{\alpha_k\}_{k\in\mathbf{N}}$ consider the definable property $P(i)$:

$$\alpha_i \in F_i \wedge F_0 \supseteq F_1 \supseteq \cdots \supseteq F_i.$$

$P(k)$ holds for all $k \in \mathbf{N}$ and so by induction in $\mathcal{M}$ it holds up to some non-standard $s<t$. By $P(i)$ we have $\alpha_i \in F_i \subseteq \bigcap_{k\in\mathbf{N}} F_k = F$, for all $i \le s$. □

3.5 Propositional computation of truth values

We shall strengthen Lemmas 3.1.2 and 3.3.3 in this section for the case of compact families. The strengthening lies in the elimination of the error term ϵ.

Theorem 3.5.1 *Let F be an L-closed compact family. Let $s \in \mathcal{M}$ be an arbitrary non-standard number.*

Let A be an L-sentence, possibly with parameters from F, and assume that A has the form:

$$\exists x_1 \forall y_1 \ldots \exists x_k \forall y_k B(x_1, y_1, \ldots, x_k, y_k)$$

with B open.

Then there are elements of F

$$\alpha_1^{i_1}, \beta_1^{i_1 j_1}, \alpha_2^{i_1 j_1, i_2}, \ldots, \beta_k^{i_1, \ldots j_k}$$

with $i_1, \ldots, j_k < s$ such that $[\![A]\!]$ is equal modulo the ideal $\mathcal{I}$ to the set, defined in $\mathcal{M}$, of all $\omega \in \Omega$ satisfying

$$\bigvee_{i_1} \bigwedge_{j_1} \cdots \bigvee_{i_k} \bigwedge_{j_k} B(\alpha_1^{i_1}(\omega), \beta_1^{i_1 j_1}(\omega), \ldots, \beta_k^{i_1, \ldots j_k}(\omega)).$$

Proof: Fix arbitrary $\epsilon > 0$, a standard rational. By Lemma 3.1.2 there are $r \in \mathbf{N}$ and elements of F:

$$\alpha_1^{i_1}, \beta_1^{i_1 j_1}, \alpha_2^{i_1 j_1, i_2}, \ldots, \beta_k^{i_1, \ldots j_k}$$

with all $i_1, \ldots, j_k < r$ such that the distance of $[\![A]\!]$ from

$$\bigvee_{i_1} \bigwedge_{j_1} \cdots [\![B(\alpha_1^{i_1}, \beta_1^{i_1 j_1}, \ldots, \beta_k^{i_1, \ldots j_k})]\!]$$

is at most ϵ. However, this $[\![\ldots]\!]$-value is equal, modulo the ideal $\mathcal{I}$, to the set of all $\omega \in \Omega$ satisfying (as evaluated in $\mathcal{M}$) the Boolean formula formed from formulas $B(\alpha_1^{i_1}(\omega), \beta_1^{i_1 j_1}(\omega), \ldots, \beta_k^{i_1, \ldots j_k}(\omega))$:

$$\bigvee_{i_1} \bigwedge_{j_1} \ldots B(\alpha_1^{i_1}(\omega), \beta_1^{i_1 j_1}(\omega), \ldots, \beta_k^{i_1, \ldots j_k}(\omega)).$$

Applying this for $\epsilon := 1/2\ell$ yields the following claim.

Claim: *For any $\ell \in \mathbf{N}$ there are $r_\ell \in \mathbf{N}$ and $\alpha_1^{i_1}, \beta_1^{i_1 j_1}, \ldots, \beta_k^{i_1, \ldots j_k} \in F$ with $i_1, \ldots, j_k < r_\ell$ such that the distance between $[\![A]\!]$ and $V_\ell / \mathcal{I}$, where V_ℓ is the set of all $\omega \in \Omega$ satisfying the formula U_ℓ:*

$$\bigvee_{i_1 < r_\ell} \bigwedge_{j_1 < r_\ell} \ldots B(\alpha_1^{i_1}(\omega), \beta_1^{i_1 j_1}(\omega), \ldots, \beta_k^{i_1, \ldots j_k}(\omega))$$

is at most $1/2\ell$.

Let $(U_w)_{w<t}$ be a non-standard extension of $(U_\ell)_{\ell\in\mathbf{N}}$ provided by $\aleph_1$-saturation. Let $V \in \mathcal{A}$ be a set such that $[\![A]\!] = V/\mathcal{I}$.

By the hypothesis that F is compact, $F = \bigcap_{k\in\mathbf{N}} F_k$ for some definable family $(F_a)_a$ (defined by an L_{all} formula H; see Definition 3.4.1), and $F_k \supseteq F_{k+1}$ for all $k \in \mathbf{N}$.

Consider the following three properties parameterized by w:

1. U_w is a formula of the form

$$\bigvee_{i_1<r}\bigwedge_{j_1<r}\dots B(\alpha_1^{i_1}(\omega),\beta_1^{i_1j_1}(\omega),\dots,\beta_k^{i_1,\dots,j_k}(\omega))$$

   for some $r < s$, and with $\alpha_1^{i_1},\beta_1^{i_1j_1},\dots,\beta_k^{i_1,\dots,j_k}$ from F_w.
2. The symmetric difference $V\Delta V_w$ has counting measure at most $1/w$.
3. $F_1 \supseteq F_2 \supseteq \cdots \supseteq F_w$.

For any standard w all three properties are met: the arity r_w is standard and so $r_w < s$, and αs and βs occurring in U_w are even from $F \subseteq F_w$. The second property is also satisfied: the counting measure of the symmetric difference $V\Delta V_w$ differs from the distance of $[\![A]\!]$ and $V_w/\mathcal{I}$ by at most an infinitesimal amount (in particular, by less than $1/2w$) and the distance itself is at most $1/2w$. The third property holds by the definition of compact families.

The properties are all definable in $\mathcal{M}$ (using formula H and V and $(U_w)_{w<t}$ as parameters) and hence by overspill must hold up to some non-standard $p < t$.

The αs and βs occurring in U_p are from F_p, which is contained in all F_k for $k \in \mathbf{N}$, hence in F. The counting measure of $V\Delta V_p$ is $< 1/p$. This is infinitesimal, so

$$[\![A]\!] = V/\mathcal{I} = V_p/\mathcal{I}.$$

□

Consider the simplest case when $A(\beta)$ has the form $\exists x B(x,\beta)$, with B open and β the only parameter from F. Then (for compact F):

$$[\![A(\beta)]\!] = \left\{\omega \in \Omega \mid \bigvee_{i<s} B(\alpha^i(\omega),\beta(\omega))\right\}/\mathcal{I}$$

for some αs from F. This does not mean that A is equal in $\mathcal{M}$ to $\bigvee_{i<s} B(\alpha^i,\beta)$, as $[\![A(\beta)]\!]$ generally is not equal (modulo $\mathcal{I}$) to the set of ωs satisfying $A(\beta(\omega))$. Also, the right-hand side is not equal to $\bigvee_{i<s}[\![B(\alpha^i,\beta)]\!]$ as taking the quotient by $\mathcal{I}$ does not commute with infinite disjunctions. Hence we should be careful in interpreting the theorem.

It is now clear that we can analogously strengthen Lemma 3.3.3.

Theorem 3.5.2 *Let F be an L-closed family that is*

1. *closed under definition by cases by open L-formulas, and*
2. *compact.*

Let A be an L-sentence of the form

$$\exists x_1 \forall y_1 \ldots \exists x_k \forall y_k B(x_1, y_1, \ldots, x_k, y_k),$$

with B open (and with parameters).

Then there are $\alpha_1, \beta_1, \ldots, \alpha_k, \beta_k \in F$ such that for all $i = 1, \ldots, k$

$$[\![\forall y_i \exists x_{i+1} \forall y_{i+1} \ldots \exists x_k \forall y_k B(\alpha_1, \beta_1, \ldots, \alpha_i, y_i, x_{i+1}, y_{i+1}, \ldots, x_k, y_k)]\!] = [\![A]\!]$$

and

$$[\![\exists x_{i+1} \forall y_{i+1} \ldots \exists x_k \forall y_k B(\alpha_1, \beta_1, \ldots, \alpha_i, \beta_i, x_{i+1}, y_{i+1}, \ldots, x_k, y_k)]\!] = [\![A]\!].$$

For compact families that we shall consider in Part III, based on various types of decision trees, Theorem 3.5.2 actually follows from Theorem 3.5.1 (formulas $B(\alpha_1^{i_1}(\omega), \beta_1^{i_1, j_1}(\omega), \ldots, \beta_k^{i_1, \ldots, j_k}(\omega))$ are then computed by decision trees and hence the whole formula too).

4

The truth in $\mathbf{N}$ and the validity in $K(F)$

In this chapter we apply a particular Skolemization to a formula in order to link its truth values in $\mathbf{N}$ and $K(F)$.

Consider an L-formula of the form $\exists y C(\overline{x}, y)$, with $\overline{x}$ a k-tuple of free variables, and C open. Let $f(\overline{x})$ be a new k-ary function symbol, and let

$$\mathrm{Sk}[\exists y C(\overline{x}, y); f]$$

be the universal closure of the formula

$$C(\overline{x}, y) \rightarrow C(\overline{x}, f(\overline{x})).$$

Clearly $\mathrm{Sk}[\exists y C(\overline{x}, y); f]$ implies (in predicate calculus) the equivalence

$$\exists y C(\overline{x}, y) \equiv C(\overline{x}, f(\overline{x})).$$

For an L-formula of the form $\forall y C(\overline{x}, y)$, again with C open and with parameters from F, we define

$$\mathrm{Sk}[\forall y C(\overline{x}, y); f]$$

to be

$$\mathrm{Sk}[\exists y \neg C(\overline{x}, y); f].$$

This axiom then implies

$$\forall y C(\overline{x}, y) \equiv C(\overline{x}, f(\overline{x})).$$

This reduction can be iterated. Let A be an L-sentence of the form

$$Q_1 x_1 \ldots Q_k x_k B(x_1, \ldots, x_k),$$

with B open and possibly with parameters. Introduce a $(k-1)$-ary function symbol f_k and axiom $\mathrm{Sk}[Q_k x_k B(x_1,\ldots,x_k);f_k]$. Then A is equivalent, modulo this axiom, to

$$Q_1x_1\ldots Q_{k-1}x_{k-1}B(x_1,\ldots,x_{k-1},f_k(x_1,\ldots,x_{k-1})).$$

Analogously introduce $(i-1)$-ary function symbols f_i and corresponding axioms, for $i=k-1,k-2,\ldots,1$ to successively reduce the quantifier complexity of the formula, the last formula

$$B(f_1,f_2(f_1),\ldots,f_k(f_1,f_2(f_1),\ldots,f_{k-1}(f_1,f_2(f_1),\ldots)))$$

(any occurrence of f_i is substituted for an occurrence of x_i – we do not show it explicitly) being a quantifier-free sentence (f_1 is a constant).

Definition 4.1 For a sentence A of the form as above, denote by

$$\mathrm{Sk}[A;f_1,\ldots,f_k]$$

the conjunction of the k universal axioms introduced in the reduction process, and by A_{Sk} the quantifier-free sentence obtained at the end.

Any k-tuple of functions $f_1,\ldots,f_k$ (of the appropriate arity) defined on **N** and satisfying $\mathrm{Sk}[A;f_1,\ldots,f_k]$ will be called a **Skolem tuple** for A.

The next lemma follows from the construction.

Lemma 4.2 *For a sentence A as above, the axiom* $\mathrm{Sk}[A;f_1,\ldots,f_k]$ *implies in predicate logic alone the equivalence*

$$A\equiv A_{\mathrm{Sk}}.$$

We aim at the following statement.

Lemma 4.3 *Assume that A is an L-sentence of the form as above that has no parameters from F. Assume also that L contains (symbols from* L_{all} *for) a Skolem tuple for A, and constants for all elements of* **N**.

Then for any L-closed family F it holds that:

$$\mathbf{N}\models A \text{ iff } [\![A]\!]=1_{\mathcal{B}}.$$

Proof: Let $f_1,\ldots,f_k$ be a Skolem tuple for A that is in L. By Lemma 1.4.2 the axiom $\mathrm{Sk}[A;f_1,\ldots,f_k]$ is $K(F)$-valid. By Lemma 4.2 then A is equivalent to A_{Sk} in both **N** and $K(F)$.

But by Lemma 1.4.2 again $\mathbf{N} \models A_{Sk}$ holds if and only if $[\![A_{\mathrm{Sk}}]\!] = 1_{\mathcal{B}}$. □

We shall use this lemma to our advantage in both directions, to pull facts from $\mathbf{N}$ to $K(F)$ and vice versa.

PART II

Second-order structures

5

Structures $K(F,G)$

We have developed some basic facts about models $K(F)$ in the preceding part, all in the first-order setting. It is often more transparent, however, to define various bounded arithmetic theories in a second-order language. This concerns, in particular, theories related to weak complexity classes like AC^0 and $AC^0(2)$ that we will be concerned with in future parts of the book.

'Second-order language' is a misnomer in this context: no second-order logic is involved. The languages of these second-order theories are many-sorted first-order languages having besides the number sort also a sort for sets (maybe also for relations or functions, when convenient). The intended range of sets consists of *bounded* sets, i.e. sets admitting an upper bound on their elements. One thinks of sets as representing binary strings and numbers as representing positions of bits or lengths of strings. In connection with computational complexity theory it is important not to require that the number sort is closed under fast growing functions.

Of course, one could use only numbers and consider sets as being coded by numbers. However, a typical subset of $[n]$ (i.e. a string of length close to n) is coded by a number proportional to 2^n. As we do not want the exponential function to be defined on all of the number sort, we would have to consider only partially defined functions. That is not difficult to do but the second-order formalism appears more convenient.

In fact, we could also have allowed unbounded sets. Although we are interested only in provability of bounded formulas, it is well-known that allowing unbounded sets does not change the strength of the theories considered as far as bounded formulas are concerned (see Buss [11], e.g. theory $V_1^0(BD)$ and others).

Paris and Wilkie [86] were first to consider a second-order version of a bounded arithmetic theory; they studied theory $I\Delta_0(R)$ extending Parikh's $I\Delta_0$ from [85]. Buss [11] then defined a variety of not so trivial second-order theories and many later authors recognized the advantages of the second-order formalism. Cook and Nguyen [30] offer an elegant and comprehensive treatment of second-order theories related to various weak computational complexity classes, as well as relevant bibliographic information.

5.1 Language L^2 and the hierarchy of bounded formulas

Let $L \subseteq L_{\text{all}}$ be any first-order language. We shall eventually consider a fairly rich language L_n (see Section 5.2) but usually the following basic language L_{PA} is adopted:

- constants 0 and 1
- functions $x+y$, $x \cdot y$
- the relation $x < y$.

Any language $L \subseteq L_{\text{all}}$ determines **language $\mathbf{L^2}$**: it is obtained from L by adding another sort of variables $X, Y, \ldots$ interpreted as bounded sets. Although the logic is always first-order we shall often call the elements of the set sort simply 'second-order'.

There is the equality symbol for sets (we consider equality as a logical symbol automatically included) which we will also denote $=$. In addition there is

- the relation $x \in X$ between numbers and sets (the membership relation).

Cook and Nguyen [30] also include the function symbol $|X|$ for a function assigning a number to a set. Its intended meaning is the least upper bound on elements of X. Buss [11] considered instead variables of the form X^t, the superscript being a first-order term bounding the elements of X. We will adopt neither formalism and instead will simply interpret the set variables as bounded sets; in our structures the axiom

$$\forall X \exists x \forall y (y \in X \rightarrow y < x)$$

will always be valid. We shall use the abbreviation $\mathbf{X < x}$ for the formula $\forall y (y \in X \rightarrow y < x)$; hence the axiom is simply

$$\forall X \exists x X < x\,.$$

As in this book we are concerned with model-theoretic constructions and not with proof-theoretic arguments, such details of the syntax are not important.

We now define a hierarchy of bounded L^2-formulas. **Bounded formulas** are those having all quantifiers (first-order or second-order) bounded, i.e. of the form: $\exists x < t$, $\forall x < t$ or $\exists X < t$, $\forall X < t$ where t is a first-order term.

The class of all bounded formulas without the set quantifiers is denoted $\Sigma_0^{1,b}$. It has an important subclass Σ_∞^b: the class of **first-order bounded L^2-formulas,** i.e. bounded L-formulas. In particular, Σ_∞^b-formulas do not have even free second-order variables.

Define simultaneously two hierarchies of classes of bounded L^2-formulas, $\Sigma_i^{1,b}$ and $\Pi_i^{1,b}$ for $i \geq 1$. These are the smallest classes of bounded formulas such that the following hold:

- $\Sigma_{i-1}^{1,b} \subseteq \Sigma_i^{1,b} \cap \Pi_i^{1,b}$.
- Both $\Sigma_i^{1,b}$ and $\Pi_i^{1,b}$ are closed under $\vee$ and $\wedge$.
- Both $\Sigma_i^{1,b}$ and $\Pi_i^{1,b}$ are closed under bounded number-quantifiers.
- If $A \in \Sigma_i^{1,b}$ then $\neg A \in \Pi_i^{1,b}$, and vice versa.
- $\Sigma_i^{1,b}$ (resp. $\Pi_i^{1,b}$) is closed under bounded existential (resp. universal) set-quantifiers.

Two particular classes of bounded L^2-formulas are **strict** $\Sigma_1^{1,b}$-formulas and **strict** $\Pi_1^{1,b}$-formulas, denoted $s\Sigma_1^{1,b}$ and $s\Pi_1^{1,b}$. These are of the form

$$QX_1 < t_1, \ldots, X_k < t_k \; \varphi$$

with Q being $\exists$ or $\forall$ respectively, and φ a $\Sigma_0^{1,b}$-formula. Cook and Nguyen [30] define the hierarchy of strict bounded formulas only, denoted Σ_i^B and Π_i^B there.

5.2 Cut $\mathcal{M}_n$, languages L_n and L_n^2

This section introduces a few objects used for the rest of the book. We first

- **fix a non-standard number** $n \in \mathcal{M}$.

The parameter n is fixed throughout the book. No particular properties of n are assumed in the following definitions but in future sections we may impose on n some additional requirement (e.g. that it is even).

Define a cut $\mathcal{M}_n \subseteq_e \mathcal{M}$ by:

$$\mathcal{M}_n := \bigcup_{t > \mathbf{N}} \{b \in \mathcal{M} \mid b \leq 2^{n^{1/t}}\} .$$

(This cut is called a **large canonical model** in Krajíček [62, 63].) Hence elements of $\mathcal{M}_n$ are numbers that are subexponential in terms of n.

Let $L_n \subseteq L_{\text{all}}$ be the language consisting of:

- all relation symbols from L_{all}, and
- those function symbols $f \in L_{\text{all}}$ such that $f''\mathcal{M}_n \subseteq \mathcal{M}_n$.

For example, if f is (a function symbol for) a polynomially bounded function

$$|f(x)| \le |x|^{O(1)}$$

then $f''\mathcal{M}_n \subseteq \mathcal{M}_n$ and $f \in L_n$.

In $\mathcal{M}$ there are no differences between numbers and $\mathcal{M}$-finite (i.e. with all elements bounded above by an element of $\mathcal{M}$) definable sets as any such set is coded by a number (and vice versa). However, in $\mathcal{M}_n$ this is no longer true: the code of a subset of $[n]$ may well be outside of $\mathcal{M}_n$.

Hence the extension of language L_n to the second-order language L_n^2 defined in Section 5.1 will be essential.

5.3 Definition of the structures

All theories under consideration have some pairing function on numbers, e.g. the usual: $\langle 0,0\rangle$ is encoded by 0 and for $i+j>0$

$$\langle i,j\rangle := \frac{(i+j)(i+j-1)}{2} + i\,.$$

Bounded functions and relations (i.e. with bounded domains and ranges) can be thus encoded by bounded sets. However, it is technically advantageous to have bounded unary functions directly as the second-order objects in the structures; sets are simply functions with values 0 and 1.

Hence we make the following definition.

Definition 5.3.1 Assume $L \subseteq L_n$. A model $K(F,G)$ for a theory in L^2 consists of an L-closed family F of functions on a sample space Ω, as in Definition 1.3.1, and a family G of some functions $\Theta \in \mathcal{M}$ assigning to $\omega \in \Omega$ a function $\Theta_\omega \in \mathcal{M}$ (it is thus $\mathcal{M}$-finite) that maps a subset $\text{dom}(\Theta_\omega)$ of $\mathcal{M}_n$ into $\mathcal{M}_n$.

The definition extends Definition 1.3.1 by defining how $\Theta \in G$ operates on F: for any $\alpha \in F$

$$\Theta(\alpha)(\omega) = \begin{cases} \Theta_\omega(\alpha(\omega)) & \text{if } \alpha(\omega) \in \mathrm{dom}(\Theta_\omega) \\ 0 & \text{otherwise.} \end{cases}$$

It is required that for all $\Theta \in G$ and all $\alpha \in F$

$$\Theta(\alpha) \in F$$

too. The value of equality is defined identically as for first-order objects:

$$[\![\Theta = \Xi]\!] := \{\omega \in \Omega \mid \theta_\omega = \Xi_\omega\}/\mathcal{I}\,.$$

Further we add two inductive clauses for second-order quantifiers:

- $[\![\exists X A(X)]\!] := \bigvee_{\Theta \in G} [\![A(\Theta)]\!]$
- $[\![\forall X A(X)]\!] := \bigwedge_{\Theta \in G} [\![A(\Theta)]\!]$.

Let us illustrate the definition by two examples. Let $\Theta, \Gamma \in G$ be (unary) functions and $\alpha, \beta \in F$. The value $[\![\Theta(\alpha) = \Gamma(\beta)]\!]$ is computed as follows: by the definition $\Theta(\alpha) = \alpha'$ and $\Gamma(\beta) = \beta'$ for some $\alpha', \beta' \in F$, we have $[\![\Theta(\alpha) = \Gamma(\beta)]\!] = [\![\alpha' = \beta']\!]$.

As a second example let us see how we represent in G a binary relation: a unary function $\Theta \in G$ and $\langle x, y\rangle$ (the pairing function on numbers, an L_n-function) define relation $\hat{\Theta}(x,y)$ by:

$$[\![\hat{\Theta}(\alpha,\beta)]\!] = [\![\Theta(\langle \alpha,\beta\rangle) = 1]\!]\,.$$

We may want to redefine Θ (by changing all its values different from 1 to 0) to get also the dual property:

$$[\![\neg\hat{\Theta}(\alpha,\beta)]\!] = [\![\Theta(\langle \alpha,\beta\rangle) = 0]\!]$$

but this might violate the property that Θ must map F into F.

In the following lemma one just needs to verify the second-order equality axioms, which is straightforward.

Lemma 5.3.2 *All L^2-sentences valid in many-sorted first-order logic with equality are valid in the two-sorted model $K(F,G)$. In particular, instances of the equality axioms*

$$\alpha = \beta \rightarrow \Theta(\alpha) = \Theta(\beta)$$

and

$$\Theta = \Xi \to \Theta(\alpha) = \Xi(\alpha)$$

are valid in $K(F,G)$.

5.4 Equality of functions, extensionality and possible generalizations

We shall now make a brief digression to discuss the issue of the definition of the equality between functions and the axiom of extensionality.

Let us restrict ourselves to sets (i.e. 0/1-valued functions) in this discussion. We write $x \in X$ in place of $X(x) = 1$. Definition 5.3.1 implies the following clauses:

- $[\![\alpha \in \Theta]\!] = \{\omega \in \Omega \mid \alpha(\omega) \in \Theta_\omega\}/\mathcal{I}$

and

- $[\![\Theta = \Xi]\!] = \{\omega \in \Omega \mid \Theta_\omega = \Xi_\omega\}/\mathcal{I}$.

With this definition we automatically obtain the validity of the relevant equality axioms in Lemma 5.3.2:

$$\alpha = \beta \to \alpha \in \Theta \equiv \beta \in \Theta$$

and

$$\Theta = \Xi \to \alpha \in \Theta \equiv \alpha \in \Xi .$$

But note that the extensionality

$$\forall X, Y \exists x (X = Y \vee x \in X \not\equiv x \in Y)$$

need not be valid in a structure $K(F,G)$ (although it will be in the models we shall define later). For example, let F consist of constants bigger than 1 while G contains two random variables, one giving constantly set $\{0\}$ and the other one giving constantly set $\{1\}$.

However, it is possible to 'arrange' extensionality by redefining the second-order equality $[\![X = Y]\!]$ as

$$[\![\Theta = \Xi]\!] := \bigwedge_\alpha [\![\alpha \in \theta \equiv \alpha \in \Xi]\!]$$

(this definition would automatically yield the second equality axioms).

In fact, one can proceed even more generally (we make the observation here although we will not take it up in this book). By taking the interpretation of sets to be also random variables in $\mathcal{M}$ we got the first equality axioms for free. However, these equality axioms are the *only* conditions we need to impose on sets in the model in order for Lemma 5.3.2 to hold. In particular, we could take for G a collection of maps

$$\mathcal{U} : F \to F$$

(we use letter $\mathcal{U}$ instead of a Greek letter to avoid confusion) such that the following inequality holds for all $\alpha, \beta \in F$:

$$[\![\alpha = \beta]\!] \leq [\![\mathcal{U}(\alpha) = \mathcal{U}(\beta)]\!] .$$

This guarantees the first equality axioms and then by defining

$$[\![\mathcal{U} = \mathcal{V}]\!] := \bigwedge_{\alpha \in F} [\![\mathcal{U}(\alpha) = \mathcal{V}(\alpha)]\!]$$

gets us the second equality axioms and the extensionality for free.

An advantage of our treatment of the second-order objects is that all results proved for $K(F)$ apply equally well to $K(F,G)$; the splitting of random variables into two families F and G is just for convenience but otherwise they are treated equally as in Definition 1.3.1. A slightly more general approach would be to have one family and a new unary predicate $Number(x)$ that defines the sort of numbers (and $\neg Number(x)$ defines the second-order sort). Our definition with separate families F and G would correspond in this more general set-up to requiring that $[\![Number(\alpha)]\!]$ is either $1_{\mathcal{B}}$ or $0_{\mathcal{B}}$ for all α.

5.5 Absoluteness of $\forall\Sigma^b_\infty$-sentences of language L_n

We now observe a simple but useful preservation property of structures $K(F,G)$.

Lemma 5.5.1 *Let A be $\forall\Sigma^b_\infty$ L_n-sentence (no parameters from F or G). Assume $K(F,G)$ is a second-order structure such that F is L_n-closed.*

Then the following holds:

$$\mathbf{N} \models A \text{ iff } [\![A]\!] = 1_{\mathcal{B}} .$$

Proof: In accordance with Lemma 4.3 it is sufficient to show that all $\forall\Sigma^b_\infty$ L_n-sentences have a Skolem tuple in L_n. This is established by induction on the number of bounded quantifiers in A, using the following claim.

Claim: *Let $A(\overline{x}, y)$ be an open L_n-formula and $t(\overline{x})$ and L_n-term. Then there is a function $f(\overline{x}) \in L_n$ such that both*

- $\forall \overline{x}\, f(\overline{x}) \leq t(\overline{x})$, *and*
- $\mathrm{Sk}[\exists y \leq t(\overline{x}, A(\overline{x}, y); f]$

are true in $\mathbf{N}$ *and valid in* $K(F, G)$.

The claim follows from the fact that L_n contains any function majorized by an L_n-term, and from Lemma 1.4.2.

□

PART III

AC^0 world

6
Theories $I\Delta_0$, $I\Delta_0(R)$ and V_1^0

In this part we shall consider models of second-order theories like $I\Delta_0(R)$ or V_1^0, related to constant-depth Frege systems and to AC^0-circuits. The common approach to the analysis of these models is Skolemization (Chapter 9), which utilizes the switching lemma from Boolean complexity (Chapter 11).

First bounded arithmetic theory $I\Delta_0$ has been defined by Parikh [85]. This is a first-order theory, a subsystem of Peano arithmetic. Its language is the L_{PA} from Section 5.1 (although sometimes the successor function is taken in place of constant 1) and the axioms are the universal closures of the following formulas:

1. $0+1=1$
2. $x+1 \neq 0$
3. $x+1=y+1 \rightarrow x=y$
4. $x \neq 0 \rightarrow \exists y(y+1=x)$
5. $x+0=x$
6. $x+(y+1)=(x+y)+1$
7. $x \cdot 0=0$
8. $x \cdot (y+1)=(x \cdot y)+x$

which comprise the so called *Robinson's Q*, together with the following set (a smaller one would suffice but there is no need to be particularly economic) of axioms about ordering (we use $x \leq y$ to abbreviate $x<y \vee x=y$):

1. $(x \leq y \wedge y \leq x) \rightarrow x=y$
2. $x<x+(y+1)$
3. $x \leq y \wedge y \leq z \rightarrow x \leq z$
4. $x \leq y \vee y \leq x$
5. $x \leq y \equiv x<y+1$

and the all-important scheme IND of bounded induction

$$A(0,\overline{y}) \wedge \forall x(A(x,\overline{y}) \rightarrow A(x+1,\overline{y})) \rightarrow \forall x A(x,\overline{y})$$

where A is a bounded L_{Pa}-formula ($\overline{y}$ are implicitly universally quantified in front of the axiom). The set of axioms excluding the induction axioms will be called PA^-. We refer the reader to Krajíček [56] for more information about $I\Delta_0$.

Theory $I\Delta_0(R)$, where $R(x,y)$ is a binary relation symbol, has been considered by Paris and Wilkie [86]. The particular choice of a binary relation symbol for R rather than unary or ternary or even a function symbol is to a large extent irrelevant. A binary relation symbol allows for an easy formalization of various interesting combinatorial principles about graphs, e.g. the pigeonhole principle.

This theory extends $I\Delta_0$ by first allowing R in the language and then accepting as an axiom any instance of the induction scheme for any bounded formula in the extended language.

One can think of $I\Delta_0(R)$ as a sort of second-order theory: there is one variable for an unknown binary relation which cannot be quantified.

Theory V_1^0 defined by Buss [11] (our version is more like $V_1^0(BD)$ of [11]) or its version V^0 in Cook and Nguyen [30] extends $I\Delta_0$ to a proper second-order theory, i.e. a theory in language L^2_{PA}.

Axioms of V_1^0 are PA^- axioms above, together with the induction axioms:

$$B(0,\overline{y}) \wedge \forall x(B(x,\overline{y}) \rightarrow B(x+1,\overline{y})) \rightarrow \forall x B(x,\overline{y}),$$

where B is a $\Sigma_0^{1,b}$-formula, and with the scheme of bounded comprehension

$$\exists X \forall y < x \; (y \in X \equiv A(y,\overline{z}))$$

also accepted for all $\Sigma_0^{1,b}$-formulas A.

Theory V_1^0 is very close to $I\Delta_0(R)$ in the following sense. If we take any model N of $I\Delta_0(R)$ and add to it as second-order objects all bounded subsets of N definable in N by a Σ_∞^b-formula in the language of $I\Delta_0(R)$ (with parameters from N) then the resulting $L^2_{PA(R)}$-structure is a model of V_1^0.

The reader may look at Krajíček [56] for more information about these theories.

In later chapters we will be concerned with the task of constructing 'interesting' structures $K(F,G)$ in which V_1^0 or stronger theories are valid. Of course, one can always achieve that by Skolemizing the theory first (for V_1^0 this can be done by introducing AC^0-computable maps from strings to strings; see Cook and Nguyen [30]) and applying the results of Chapter 4. Using the witnessing theorems from Chapter 3 for the analysis of the resulting model is then quite analogous to using the Herbrand theorem on the Skolemized theory, and the model seems consequently quite dull. In other words, the qualification 'interesting' is essential.

7
Shallow Boolean decision tree model

In this chapter we construct a rudimentary example of a second-order structure $K(F_{\text{rud}}, G_{\text{rud}})$, and we shall establish its basic properties. The particular models used in later chapters will all be related to this rudimentary example. In particular, it will be clear that the basic properties of $K(F_{\text{rud}}, G_{\text{rud}})$ established here hold also (and for exactly the same reasons) in these models. This will allow us to concentrate on establishing specific properties of each of these particular models.

7.1 Family F_{rud}

This whole section is one long definition; we shall not split it into a sequence of smaller definitions with their own numbering. Recall the definition of the cut $\mathcal{M}_n$ from Section 5.2, and use the standard notation $[n] := \{1,\ldots,n\}$.

The sample space is

$$\Omega := \{\omega \in \mathcal{M} \mid \omega \subseteq [n]\}.$$

To define the family F_{rud} we have to define first an auxiliary concept of a labeled tree. A **decision tree** T is defined as usual: it is a finite binary tree whose inner nodes (i.e. non-leafs) are labeled by queries of the form $i \in_? \omega$, for $i \in [n]$. The two outgoing edges from such a node are labeled yes and no, respectively. The **depth of a tree**, denoted $\text{dp}(T)$, is the length of the longest path (i.e. the number of edges on it) from the root to a leaf. A depth-0 tree thus consists of just one node, a leaf. A **labeling** of T is a map ℓ assigning to leafs of T values in $\mathcal{M}_n$.

Let (T,ℓ) be a labeled tree. Any $\omega \in \Omega$ determines a unique path through T ending in a leaf that we shall denote $T(\omega)$. In this way (T,ℓ) defines a function

$$\omega \in \Omega \to \ell(T(\omega)) \in \mathcal{M}_n.$$

Family F_{rud} consists of all functions

$$\alpha : \Omega \rightarrow \mathcal{M}_n$$

from $\mathcal{M}$ such that there is labeled tree $(T, \ell) \in \mathcal{M}$ such that

- $\alpha(\omega) = \ell(T(\omega))$, for all $\omega \in \Omega$, and
- the depth $\text{dp}(T)$ of T satisfies $\text{dp}(T) < n^{1/t}$, for some $t > \mathbf{N}$.

In particular, depth-0 trees labeled by elements of $\mathcal{M}_n$ represent in F_{rud} constants from $\mathcal{M}_n$.

7.2 Family G_{rud}

The definition of G_{rud} (Definition 7.2.3) requires some preliminary notions.

Definition 7.2.1 Let $m \in \mathcal{M}_n$ and let $\hat{\beta} = (\beta_0, \ldots, \beta_{m-1}) \in \mathcal{M}$ be an m-tuple of elements of F_{rud}.

For any $\alpha \in F_{\text{rud}}$ define $\hat{\beta}(\alpha) : \Omega \rightarrow \mathcal{M}_n$ by

$$\hat{\beta}(\alpha)(\omega) = \begin{cases} \beta_{\alpha(\omega)}(\omega) & \text{if } \alpha(\omega) < m \\ 0 & \text{otherwise.} \end{cases}$$

Lemma 7.2.2 *Let $m \in \mathcal{M}_n$ and let $\hat{\beta} = (\beta_0, \ldots, \beta_{m-1}) \in \mathcal{M}$ be an m-tuple of elements of F_{rud}.*

Then function $\hat{\beta}$ from Definition 7.2.1 maps F_{rud} into F_{rud}.

Proof: As $\hat{\beta} \in \mathcal{M}$, the induction in $\mathcal{M}$ implies that there is a maximal depth of a tree computing one of β_i, $i < m$. This depth is therefore bounded above by n raised to an infinitesimal. The depth of the tree computing any $\alpha \in F_{\text{rud}}$ is bounded by a term of the same form, and hence so is the depth of the tree computing $\hat{\beta}(\alpha)$. □

Definition 7.2.3 Family G_{rud} consists of all random variables Θ computed by some $\hat{\beta} = (\beta_0, \ldots, \beta_{m-1})$, where $m \in \mathcal{M}_n$ and $\hat{\beta} \in \mathcal{M}$ is a m-tuple of elements of F_{rud}.

In other words, for Θ computed by $\hat{\beta}$ the function Θ_ω has the graph $\{(i, \beta_i(\omega)) \mid i < m\}$.

7.3 Properties of F_{rud} and G_{rud}

Let $L_n(F_{\text{rud}}, G_{\text{rud}})$ be the language L_n augmented by names for elements of F_{rud} and G_{rud}.

Lemma 7.3.1 *F_{rud} is an L_n-closed family. It is closed under definition by open $L_n(F_{\text{rud}}, G_{\text{rud}})$-formulas, and it is compact. Also G_{rud} is compact.*

Proof: Let $f(x,y)$ be a binary function symbol from L_n (functions of higher arity are treated analogously). Let $\alpha, \beta \in F_{\text{rud}}$ be two random variables computed by trees (T_α, ℓ_α) and (T_β, ℓ_β) respectively. The element $f(\alpha, \beta)$ is computed by (T, ℓ) where T is obtained from T_α by replacing each leaf by a copy of T_β, and where the label $\ell(T(\omega))$ is defined to be

$$\ell(T(\omega)) := f(\ell_\alpha(T_\alpha(\omega)), \ell_\beta(T_\beta(\omega))).$$

This is a sound definition as both $T_\alpha(\omega)$ and $T_\beta(\omega)$ are determined by $T(\omega)$, and $\text{dp}(T)$ clearly obeys the bounds imposed by the definition of F_{rud} if $\text{dp}(T_\alpha)$ and $\text{dp}(T_\beta)$ do. This verifies the L_n-closedness of F_{rud}.

The truth values of atomic sentences

$$R(\alpha_1, \ldots, \alpha_k) \quad \text{or} \quad f(\alpha_1, \ldots, \alpha_k) = g(\alpha_1, \ldots, \alpha_k),$$

for $R, f, g \in L_n$, can be clearly computed by decision trees composed from the labeled decision trees computing the individual α_is. The depth of the tree is the sum of the depths of the individual trees and hence obeys the required bounds.

The truth value of an atomic sentence of the form $\Theta(\alpha) = \gamma$ is computed by combining the tree T extending T_α by adding trees computing Θ given by $\hat{\beta} = (\beta_0, \ldots, \beta_{m-1})$ (at a leaf of T_α labeled by i compute β_i and label the new leafs accordingly), with the tree computing γ. Hence it is again computed by a tree from F_{rud}. It follows that F_{rud} is closed under definition by cases by open $L_n(F_{\text{rud}}, G_{\text{rud}})$-formulas.

Finally, to verify the compactness of F_{rud}, define sets F_a to consists of those functions

$$\alpha \,:\, \Omega \to \{b \in \mathcal{M} \mid b < 2^{n^{1/a}}\}$$

computed by a tree of depth less than $n^{1/a}$. So $F_{\text{rud}} = \bigcap_{k \in \mathbf{N}} F_k$. Similarly, sets G_a consisting of functions computed by $\hat{\beta} = (\beta_0, \ldots, \beta_{m-1})$, with $m < 2^{n^{1/a}}$ and all $\beta_i \in F_a$, certify that G_{rud} is compact too. □

8
Open comprehension and open induction

Arranging that some modicum of induction is valid in a model is a key requirement. First, the theories in question usually have induction as their principal axiom. Second, some induction is also needed in order to deduce lower bounds for propositional proof system (see Part V). Generally, induction for a wider class of formulas yields lower bounds for a stronger proof system.

In this chapter we start investigating how to arrange induction in a model.

8.1 The $\langle\langle \ldots \rangle\rangle$ notation

In analyzing practically all models we will compute the truth values of various open sentences by considering the subset of the sample space where the instance of the sentence is true. This has actually already been done in Chapter 3 but now a need for some abbreviated notation will become especially pressing. We introduce it now.

Let $K(F,G)$ be an arbitrary L^2-structure, for any $L \subseteq L_{all}$.

Definition 8.1.1 For an L^2-sentence $B(\alpha_1,\ldots,\alpha_k,\Theta_1,\ldots,\Theta_\ell)$, with all parameters from F and G shown, define:

$$\langle\langle B(\alpha_1,\ldots,\alpha_k,\Theta_1,\ldots,\Theta_\ell)\rangle\rangle :=$$

$$\{\omega \in \Omega \mid B(\alpha_1(\omega),\ldots,\alpha_k(\omega),(\Theta_1)_\omega,\ldots,(\Theta_\ell)_\omega)\}.$$

The next lemma follows from the definition of the Boolean evaluation $[[\ldots]]$ and the fact that taking quotient modulo the ideal $\mathcal{I}$ commutes with finite Boolean combinations.

Lemma 8.1.2 *Let $K(F,G)$ be an arbitrary L^2-structure and assume B is an open L^2-sentence with parameters from F or G. Then:*

$$[\![B]\!] \;=\; \langle\!\langle B\rangle\!\rangle/\mathcal{I}.$$

Note that for a general formula there is no a priori relation between these two Boolean values. For example, let A be a sentence of the form $\exists yR(\alpha,y)$, where $\alpha \in F$ and $R(x,y)$ is a relation from L such that $\forall x\exists yR(x,y)$ holds true in $\mathbf{N}$ but no $\beta \in F$ finds a witness for $\exists yR(\alpha(\omega),y)$ with a non-infinitesimal probability. Then $[\![A]\!] = 0_{\mathcal{B}}$ while $\langle\!\langle A\rangle\!\rangle = \Omega = 1_{\mathcal{A}}$.

8.2 Open comprehension in $K(F_{\mathrm{rud}}, G_{\mathrm{rud}})$

We note in this section that comprehension for open formulas is a corollary of two simple technical lemmas.

Lemma 8.2.1 *Let $A(y)$ be an open $L_n^2(F_{\mathrm{rud}}, G_{\mathrm{rud}})$-formula with y the only free variable.*

Then there is $t > \mathbf{N}$ such that for all $i \in \mathcal{M}_n$ the membership in $\langle\!\langle A(i)\rangle\!\rangle$ is computed by a tree with the leafs labeled by 1 (for yes*) or 0 (for* no*) of the depth bounded above by $n^{1/t}$. In particular, the membership is computed by an element of F_{rud}.*

Proof: Instances $t(i)$ of any term $t(y)$ can be computed by trees satisfying the restrictions; this is verified by induction on the complexity of the term. Equations among terms and their Boolean combinations can be decided when all terms involved are computed; this composition of finitely many trees again obeys the restrictions. □

Taking for m an element of $\mathcal{M}_n$ that is bigger than all labels of α yields the following statement.

Lemma 8.2.2 *Any $\alpha \in F_{\mathrm{rud}}$ is bounded above by some $m \in \mathcal{M}$ in the sense that $[\![\alpha < m]\!] = 1_{\mathcal{B}}$. In fact, one can find m such that even $\langle\!\langle \alpha < m\rangle\!\rangle = \Omega$.*

By the lemma it suffices to consider comprehension of the form

$$\exists X \forall y < m\; (y \in X \equiv A(y))$$

with A allowing parameters. Taking as a witness for X the element of G_{rud} computed by $\hat{\beta} = (\beta_0, \ldots, \beta_{m-1})$, where trees β_i are the trees provided by Lemma 8.2.1 computing the membership in $\langle\!\langle A(i)\rangle\!\rangle$, yields the open comprehension.

Corollary 8.2.3 *The axiom of comprehension is valid in $K(F_{\mathrm{rud}}, G_{\mathrm{rud}})$ for all open formulas.*

8.3 Open induction in $K(F_{\mathrm{rud}}, G_{\mathrm{rud}})$

Assume we have an open formula $A(x)$ with parameters from F_{rud} and G_{rud} that we do not show explicitly. We want to show that the induction axiom:

$$\neg A(0) \lor A(m) \lor \exists x < m\, (A(x) \land \neg A(x+1))$$

is valid in $K(F_{\mathrm{rud}}, G_{\mathrm{rud}})$. Here $m \in \mathcal{M}_n$; by Lemma 8.2.2 without loss of generality we may consider induction up to an element of $\mathcal{M}_n$ rather than up to an element of F_{rud}.

We may also assume without loss of generality that $[\![A(0)]\!] = 1_{\mathcal{B}}$ and $[\![A(m)]\!] = 0_{\mathcal{B}}$ and, in fact, that

$$\langle\langle A(0)\rangle\rangle = \Omega \text{ and } \langle\langle A(m)\rangle\rangle = \emptyset.$$

This can be achieved by replacing $A(x)$ by another open formula

$$A'(x) := x = 0 \lor (x < m \land A(x))$$

if necessary, induction for which is equivalent to the induction for the original formula (over a few axioms about linear ordering).

Lemma 8.3.1 *Assume that there is $\alpha \in F_{\mathrm{rud}}$ such that for all $\omega \in \Omega$ if $\alpha(\omega) = i$ then $i < m$ and*

$$\omega \in \langle\langle A(i)\rangle\rangle \setminus \langle\langle A(i+1)\rangle\rangle.$$

Then

$$[\![\alpha < m \land A(\alpha) \land \neg A(\alpha+1)]\!] = 1_{\mathcal{B}},$$

i.e. α witnesses the induction axiom for $A(x)$.

Proof: Clearly $[\![\alpha < m]\!] = 1_{\mathcal{B}}$. By the hypothesis

$$\langle\langle A(\alpha)\rangle\rangle = \bigcup_{i<m} (\{\omega \in \Omega \mid \alpha(\omega) = i\} \cap \langle\langle A(i)\rangle\rangle)$$

$$= \bigcup_{i<m} \{\omega \in \Omega \mid \alpha(\omega) = i\} = \Omega.$$

So $[\![A(\alpha)]\!] = 1_{\mathcal{B}}$, and similarly one shows that $[\![\neg A(\alpha+1)]\!] = 1_{\mathcal{B}}$. □

It is easy to define by binary search a function $\alpha : \Omega \to M$, $\alpha \in \mathcal{M}$, that satisfies the hypothesis of the preceding lemma. Namely, given $\omega \in \Omega$, construct pairs (i_t, j_t) such that $(i_0, j_0) = (0, m)$ and each (i_{t+1}, j_{t+1}) is either the second or the first half of (i_t, j_t), depending on whether or not the midpoint $k := \lfloor \frac{i_t + j_t}{2} \rfloor$ satisfies $\omega \in \langle\langle A(k) \rangle\rangle$.

The issue remains whether this α is an element of F_{rud}. By Lemma 8.2.1 there is a bound $d = n^{1/t}, t > \mathbf{N}$, on the depth of trees computing the membership in all $\langle\langle A(k) \rangle\rangle$. It is then easy to see that α is computed by a tree of depth $\leq d \cdot \log(m)$. As $m \in \mathcal{M}_n$, $\log(m)$ is also bounded by n raised to an infinitesimal and so is $d \cdot \log(m)$. Hence $\alpha \in F_{\text{rud}}$ and we have the following lemma.

Lemma 8.3.2 *Open induction is valid in $K(F_{\text{rud}}, G_{\text{rud}})$.*

8.4 Short open induction

The ordinary induction axioms, as in Section 8.3, talk about the validity of induction on intervals $[0, m]$, where $m \in \mathcal{M}_n$ is arbitrary. If one considers induction for a class of formulas closed under Boolean combinations and bounded quantification (as is, for example, the class $\Sigma_0^{1,b}$) then, in fact, the general induction already follows from a 'short induction', an induction on intervals $[0, \log(m)]$. This is shown by Solovay's method of shortening cuts (see Hájek and Pudlák [38] or Krajíček [56]) that we shall describe for completeness below.

Assume that

$$A(0) \wedge \forall x(A(x) \to A(x+1))$$

holds but $\neg A(m)$.

Define $B(x) := \forall y \leq x A(y)$. It follows (again using only a few ordering axioms) that B also satisfies

$$B(0) \wedge \forall x(B(x) \to B(x+1)) \wedge \neg B(m)$$

but in addition it is closed downwards:

$$(B(x) \wedge y \leq x) \to B(y),$$

i.e. B is a cut.

Now put

$$C(x) := \forall y < m(B(y) \to B(x+y)).$$

It is easy to verify that C is also a cut and $\neg C(m)$ but C is closed under more functions:

$$(x < m \wedge C(x)) \to (C(2x) \wedge C(2x+1)).$$

To see this, assume $C(x)$. To entail $C(2x)$ we need to verify that

$$\forall y < m(B(y) \to B(2x+y)).$$

For any $y < m$, $B(y) \wedge C(x)$ imply $B(x+y)$, which again with $C(x)$ implies (as necessarily $x+y < m$) $B(2x+y)$. The closure of C under $x \to 2x+1$ follows as it is a cut.

Now define the last formula in this sequence of definitions:

$$D(x) := \exists y < m(C(y) \wedge |y| = x),$$

i.e. D consists of the lengths of elements in C. Then clearly $D(0)$ and $\neg D(\log(m))$ hold. D is also a cut (closed under the successor function) because of the closure properties of C.

Putting it all together: if induction for A fails on $[0,m]$ then induction for D fails on $[0,\log(m)]$.

Now we will observe that it is easy to witness this 'short' induction for open formulas in $K(F_{\text{rud}}, G_{\text{rud}})$.

Lemma 8.4.1 *Assume $D(x)$ is an open formula with parameters from F_{rud} or G_{rud} and let $m \in \mathcal{M}_n$.*

Then there is $\alpha \in F_{\text{rud}}$ such that

$$[\![(D(0) \wedge \neg D(\log(m))) \to (\alpha < \log(m) \wedge D(\alpha) \wedge \neg D(\alpha+1))]\!] = 1_{\mathcal{B}}.$$

Proof: For $i = 0, 1, \ldots, \log(m)$ let β_i be a tree computing the membership into $\langle\langle D(i) \rangle\rangle$ (use Lemma 8.2.1). Putting all these $\log(m)+1$ trees below one another will produce a tree of depth still bounded above by n raised to an infinitesimal. Using the tree we can define a random variable $\alpha \in F_{\text{rud}}$ such that, for all $\omega \in \Omega$, $\alpha(\omega)$ equals the minimal $i < \log(m)$ such that $\omega \in \langle\langle D(i) \rangle\rangle$.

It is easy to verify, using Lemma 8.1.2, that α satisfies the requirement. □

Formula D resulting from Solovay's construction is not open. Hence this lemma is useful only in the presence of an elimination of bounded quantifiers.

9

Comprehension and induction via quantifier elimination: a general reduction

The last chapter showed that open comprehension and open induction are fairly easy to verify in the structure $K(F_{\text{rud}}, G_{\text{rud}})$, and that will be the same for most of the models we shall consider. A way to arrange comprehension and induction for more complex formulas is quantifier elimination.

In this chapter we present a few well-known definitions adapted to structures $K(F,G)$ and to the way they will be used later, and we show that they are enough for the general reduction of comprehension and induction for $\Sigma_0^{1,b}$-formulas to their open cases.

9.1 Bounded quantifier elimination

The qualification 'bounded' in the title of the section has a double meaning: we will eliminate bounded first-order quantifiers and on bounded domains only.

Definition 9.1.1 Let $K(F,G)$ be an L_n^2-structure. Let $B(x_1,\ldots,x_k)$ be a bounded first-order $L_n^2(F,G)$-formula, i.e. a Σ_∞^b-formula, possibly with parameters from F or G.

We say that B admits bounded quantifier elimination in $K(F,G)$ if for all $m \in \mathcal{M}_n$ there is an open $L_n^2(F,G)$-formula $A(x_1,\ldots,x_k)$ such that

$$[\![\forall x_1,\ldots,x_k < m\, B(x_1,\ldots,x_k) \equiv A(x_1,\ldots,x_k)]\!] = 1_{\mathcal{B}}.$$

Note that formula A may have other parameters from F and G over those present in B.

9.2 Skolem functions in $K(F,G)$ and quantifier elimination

Theorem 3.5.2 already provides a form of quantifier elimination for the rudimentary model (using Lemma 7.3.1). Namely, for a formula $A(x)$ of the form $Q_1y_1 \dots Q_ky_kA(x,\overline{y})$ with A open and any $\alpha \in F_{\text{rud}}$, there are $\gamma_1,\dots,\gamma_k \in F_{\text{rud}}$ such that

$$[\![B(\alpha) \equiv A(\alpha,\gamma_1,\dots,\gamma_k)]\!] = 1_{\mathcal{B}}.$$

In fact, we could allow second-order quantifiers in B too, by Lemma 7.3.1.

However, that falls short of what is required in Definition 9.1.1. We need a form of quantifier elimination that describes explicitly the dependence of γ_is on α in order to get the parametric version required in Definition 9.1.1.

Definition 9.2.1 Let $B(\overline{x},y)$ be an open first-order $L_n^2(F,G)$-formula, $\overline{x} = (x_1,\dots,x_k)$. Let $m \in \mathcal{M}_n$.

For any $\Theta \in G$ satisfying for all $\alpha_i \in F$, $i \le k$, the equation

$$[\![\bigwedge_{i\le k}\alpha_i < m \rightarrow [\exists y < mB(\overline{\alpha},y) \rightarrow (\Theta(\overline{\alpha}) < m \wedge B(\overline{\alpha},\Theta(\overline{\alpha})))]]\!] = 1_{\mathcal{B}}$$

is called a Skolem function for $\exists yB(\overline{x},y)$ on m.

The phrase ‘Skolem function on m’ allows us to abbreviate the formula; otherwise it would have to be written in a longer form as:

$$\bigwedge_{i\le k} x_i < m \rightarrow \exists y < mB(\overline{x},y).$$

Definition 9.2.2 A structure $K(F,G)$ is Skolem closed iff for all open first-order $L_n^2(F,G)$-formulas $B(\overline{x},y)$ and for all $m \in \mathcal{M}_n$ there is a $\Theta \in G$ that is a Skolem function for $\exists yB(\overline{x},y)$ on m.

The following theorem is a straightforward analogue to the standard derivation of quantifier elimination from the existence of Skolem functions; see also Chapter 4.

Theorem 9.2.3 *Assume that an L_n^2-structure $K(F,G)$ is Skolem closed. Then any Σ_∞^b $L_n^2(F,G)$-formula admits bounded quantifier elimination in $K(F,G)$.*

9.3 Comprehension and induction for $\Sigma_0^{1,b}$-formulas

Here we just summarize for later applications the general reduction of general comprehension and induction to the open case via Skolemization.

Theorem 9.3.1 *Let $K(F,G)$ be an L^2_n-structure. Assume the following:*

1. *Open comprehension is valid in $K(F,G)$.*
2. *Open induction is valid in $K(F,G)$.*
3. *$K(F,G)$ admits bounded quantifier elimination for all Σ^b_∞ $L^2_n(F,G)$-formulas.*

Then comprehension and induction for all $\Sigma^{1,b}_0$ $L^2_n(F,G)$-formulas is valid in $K(F,G)$. In particular, theory V^0_1 is valid in $K(F,G)$.

The conclusion holds also if condition 3 is replaced by 3′:

3′. $K(F,G)$ is Skolem closed.

Proof: Assume we have a $\Sigma^{1,b}_0$ $L^2_n(F,G)$-formula for which we want to verify, say, the induction axiom up to $m \in \mathcal{M}_n$. The free variables of the formula excluding the induction variable are universally quantified in front of the axiom. Hence we substitute for these parameters any elements of F or G respectively, and verify the axiom for all theses instances.

Note that after the substitution we get a bounded first-order $L^2_n(F,G)$-formula $B(x)$. By the bounded quantifier elimination there is an open $L^2_n(F,G)$-formula $A(x)$ such that

$$[\![\forall x < m+1\ B(x) \equiv A(x)]\!] = 1_{\mathcal{B}}.$$

By the validity of open induction we also have

$$[\![\neg A(0) \lor A(m) \lor \exists x < m(A(x) \land \neg A(x+1))]\!] = 1_{\mathcal{B}}.$$

These two sentences (together with a few valid axioms about ordering) in predicate logic alone imply the induction for $B(x)$ on $[0,m]$, and the provability in predicate logic preserves the truth value $1_{\mathcal{B}}$.

Comprehension axioms are verified analogously. This proves the first part of the theorem. The second part follows from Theorem 9.2.3. □

10

Skolem functions, switching lemma and the tree model

In this chapter we modify the rudimentary model based on shallow decision trees to a model using deep trees (we leave the adjective 'deep' out), in order to get Skolem functions into a family G and thus provide bounded quantifier elimination for $\Sigma_0^{1,b}$-formulas (in the next chapter).

10.1 Switching lemma

We start by recalling the classical topic of partial random restrictions and the switching lemma. This was invented by Furst, Saxe and Sipser [34] and Ajtai [1], and strengthened by Yao [108] and Hastad[40], and later generalized for the purpose of proof complexity in Krajíček, Pudlák and Woods [79] and in Pitassi, Beame and Impagliazzo [88] and elsewhere (see Chapter 21 or Pudlák [89]).

The prototype form of the switching lemma is the one due to Hastad [40]. We state it in a form useful for our purposes.

Definition 10.1.1 Let $X \subseteq [n]$ be arbitrary. For such X let D_X be the set of all partial truth assignments $\rho : X \to \{0,1\}$ with domain $\mathrm{dom}(\rho) = X$. For $\rho \in D_X$ define

$$E_\rho := \{\omega : [n] \to \{0,1\} \mid \rho \subseteq \omega\}.$$

Note that if we identify the set of $\omega : [n] \to \{0,1\}$ with $\{0,1\}^n$ (i.e. with the sample space of the rudimentary model), E_ρ defines a subset of the sample space.

Theorem 10.1.2 (Hastad's switching lemma [40]) *Let $1 \leq k, \ell$ and $0 \leq m < n$ be arbitrary parameters. Suppose propositional formula $\varphi(x_1,\ldots,x_n)$ is in a*

k-DNF form, i.e. it is a disjunction of terms (conjunctions of literals) of size at most k.

Pick a subset $X \subseteq [n]$ uniformly at random from all subsets of $[n]$ of size m and then pick uniformly at random a partial restriction $\rho \in D_X$.

The probability that the restricted formula $\varphi \downarrow \rho$ (i.e. φ partially evaluated by ρ) cannot be computed by a decision tree (over variables x_i with $i \notin X$) of depth at most $k \cdot \ell$ is bounded above by:

$$\left(\frac{10k(n-m)}{n}\right)^{\ell}.$$

As pointed out in Section 9.2, Theorem 3.5.2 already provides a form of quantifier elimination. In particular, for any formula of the form $\exists y < mB(\alpha, y)$ there is $\beta \in F_{\mathrm{rud}}$ such that

$$[\![\exists y < mB(\alpha,y)]\!] = [\![\beta < m \wedge B(\alpha,\beta)]\!].$$

But the method of the proof of Theorem 3.5.2, utilizing the ccc of $\mathcal{B}$ and the compactness of F_{rud}, does not give information about the dependence of β on α and, in particular, does not yield a Skolem function $\Theta \in G_{\mathrm{rud}}$ that would satisfy

$$[\![\exists y < mB(\alpha,y)]\!] = [\![\Theta(\alpha) < m \wedge B(\alpha,\Theta(\alpha))]\!]$$

for all $\alpha \in F_{\mathrm{rud}}$.

We will now describe how the switching lemma can be used to construct such a Skolem function. The actual implementation of the idea requires a modification of the sample space as well as of the families F_{rud} and G_{rud}. Thus we describe the idea only informally here and develop it formally in Chapter 11, after describing the new structure $K(F_{\mathrm{tree}}, G_{\mathrm{tree}})$ in Section 10.2.

Take an open $L_n^2(F_{\mathrm{rud}}, G_{\mathrm{rud}})$-formula $B(x,y)$ and some bound $m \in \mathcal{M}_n$. Let $\alpha, \beta \in F_{\mathrm{rud}}$ and assume:

$$[\![\exists y < mB(\alpha,y)]\!] = [\![\beta < m \wedge B(\alpha,\beta)]\!].$$

Now let us compute using the $\langle\!\langle \dots \rangle\!\rangle$ notation.

$$\begin{aligned}\langle\!\langle \beta < m \wedge B(\alpha,\beta)\rangle\!\rangle &= \bigcup_{j<m} \langle\!\langle \beta = j \wedge B(\alpha,\beta)\rangle\!\rangle \\ &= \bigcup_{j<m} \langle\!\langle \beta = j \wedge B(\alpha,j)\rangle\!\rangle \subseteq \bigcup_{j<m} \langle\!\langle B(\alpha,j)\rangle\!\rangle.\end{aligned}$$

Here $j < m$ range over elements of $\mathcal{M}$, not $F_{\rm rud}$. But note that the last expression satisfies:

$$\langle\langle \exists y < mB(\alpha, y)\rangle\rangle = \bigcup_{j<m} \langle\langle B(\alpha, j)\rangle\rangle.$$

Hence a way to find β witnessing the quantifier is to find a decision tree that on input ω finds some $j < m$ such that $\omega \in \langle\langle B(\alpha, j)\rangle\rangle$, if it exists. But if we could decide for any $1 \leq u < v < m$ by a tree from $F_{\rm rud}$ of depth d independent of u, v any disjunction

$$\bigvee_{u \leq j \leq v} \omega \in \langle\langle B(\alpha, j)\rangle\rangle$$

we could use a binary search to create a depth $d \cdot \log(m)$ tree finding some j.

By Lemma 8.2.1 formula $\omega \in \langle\langle B(\alpha, j)\rangle\rangle$ is computed by a tree β_j from $F_{\rm rud}$ of depth $n^{1/t}$ independent of j. Thus the statement $\omega \in \langle\langle B(\alpha, j)\rangle\rangle$ is equivalent to a disjunction of terms of size $n^{1/t}$: the terms correspond to paths in the tree β_j that end in a leaf labeled 1 (i.e. yes). Consequently, each disjunction

$$\bigvee_{u \leq j \leq v} \omega \in \langle\langle B(\alpha, j)\rangle\rangle$$

is equivalent to a k-DNF, with $k := n^{1/t}$. It is here that the switching lemma is applied and provides us with a decision tree of depth $n^{2/t}$ computing the validity of the disjunction.

To construct the witness from α by a Skolem function needs some additional tricks that we will postpone until Chapter 11. There are two conceptual problems caused by the switching lemma with regard to the rudimentary shallow tree model $K(F_{\rm rud}, G_{\rm rud})$. The first issue is that the switching lemma guarantees the existence of the desired decision tree only after a partial restriction on a fairly large domain (we will need $m \geq n - n^{1-\epsilon}$). Consequently, if we wanted to interpret the construction on a subset of the sample space $\{0,1\}^n$ satisfying the restriction, this subset would have an infinitesimal counting measure in Ω and hence contribute nothing to the truth values in question.

The second issue is that, although the probability of the restriction in the switching lemma failing is very small, it is non-zero and hence one restriction will not work for all formulas at once. In other words, we could construct the Skolem function in this way only for some sets of formulas, not for all.

It turns out that both issues can be successfully addressed by defining a different sample space in the next section.

10.2 Tree model $K(F_{\text{tree}}, G_{\text{tree}})$

This section is devoted to the definition of the new sample space Ω_{tree} and the new structure $K(F_{\text{tree}}, G_{\text{tree}})$. The definition of Ω_{tree} may look a bit unfriendly at first but this is compensated for later by the smooth way it works.

For the definition we use as an auxiliary parameter a function assigning to any $k \in \mathcal{M}$ an $\mathcal{M}$-rational δ_k by the formula:

$$\delta_k := \frac{1+2^{k+1}}{2(1+2^k)}.$$

The only two properties of this function we actually use are:

(i) $0 < \delta_k < 1$, for all $k \geq 0$,
(ii) $\delta_0 \cdot \delta_1 \cdot \dots \cdot \delta_k > \frac{1}{2}$, for all $k \geq 0$.

Any function with these two properties would serve equally well.

Definition 10.2.1 For $k \geq 0$ define inductively a **level** k **tree**, or briefly a k-tree, as follows:

- A 0-tree is any decision tree T such that there is $X \subseteq [n]$ and the following hold:
  - $|X| = n - n^{\delta_0}$ (rounded to the nearest integer).
  - T on each path queries, in the natural order on $[n]$, the membership of all elements of X (in input ω).

  Set X is called the **support** of each leaf in T.
- A $(k+1)$-tree is any tree T for which there exists a k-tree S such that the following holds:
  If $X_\ell \subseteq [n]$ are the supports of leafs ℓ of S then there are subsets $Y_\ell \subseteq [n] \setminus X_\ell$ such that
  - $|Y_\ell| = (n - |X_\ell|) - (n - |X_\ell|)^{\delta_{k+1}}$,
  - at each leaf ℓ T extends S by querying on each path, in the natural order on $[n]$, the membership of all elements of Y_ℓ.

  The support of any new leaf ℓ' in T that is below an old leaf ℓ of S is $X_\ell \cup Y_\ell$.

Let Tree_k denote the set of all k-trees.

A partial ordering $\preceq$ on these trees is defined as follows: for a k-tree T and an ℓ-tree it holds $T \preceq S$ iff T the initial part of S.

Note that if T and S are both k-trees then $T \preceq S$ iff $T = S$, i.e. different trees of the same level are $\preceq$-incomparable.

Lemma 10.2.2 *For any $k \geq 0$ (in $\mathcal{M}$) the depth of a k-tree is less than $n - n^{1/2}$.*

Proof: The depth d_k of a k-tree is $n - n^s$ where $s = \delta_0 \cdot \delta_1 \cdot \cdots \cdot \delta_k = \frac{1+2^{k+1}}{2^{k+2}} > \frac{1}{2}$. □

For the next definition we need a fixed non-standard number h as a parameter. We could select h in some canonical way, e.g. $h = n$. But taking a new symbol h for it seems to stress more that its role is auxiliary and its value is, as long as it is non-standard, irrelevant. We will not refer to h in the forthcoming notation.

Definition 10.2.3 The sample space Ω_{tree} is

$$\bigcup_{T \in \text{Tree}_h} \{T\} \times \{0,1\}^n.$$

Now comes the key definition of the family of random variables F_{tree}. Recall from the proof of Lemma 10.2.2 that the depth d_k of k-trees is

$$d_k := n - n^{\frac{1+2^{k+1}}{2^{k+2}}} < n - n^{\frac{1}{2}}.$$

Definition 10.2.4 The family F_{tree} consists of all functions $\alpha \in \mathcal{M}$,

$$\alpha : \Omega_{\text{tree}} \to \mathcal{M}_n,$$

for which there exists $k \in \mathbf{N}$, $t > \mathbf{N}$ and a tuple $(S_Z)_{Z \in \text{Tree}_k} \in \mathcal{M}$ of labelled trees such that the following hold.

- Each S_Z extends Z by adding to leafs of Z some subtrees.
- The depth of all S_Z in the tuple is bounded by $d_k + n^{1/t}$.
- The value of α on sample $(T, \omega) \in \Omega_{\text{tree}}$ is computed by S_Z at ω, for the unique $Z \preceq T$.

The parameter k is called the **level** of α.

Note, in particular, that any constant from $\mathcal{M}_n$ is represented by any tuple of such trees whose leafs (of all trees) are all labeled by the constant.

In this definition we quantify over standard k and non-standard t, and hence the family F_{tree} is not definable in $\mathcal{M}$. While the definability of the family of random variables is not required by the method, the definability of the sample space is. For this reason we used the auxiliary parameter h and the set Tree_h in place of undefinable $\bigcup_{k \in \mathbf{N}} \text{Tree}_k$.

We need a simple technical observation.

Lemma 10.2.5 *Let $m \in \mathcal{M}_n$ and let $\hat{\beta} = (\beta_0, \ldots, \beta_{m-1}) \in \mathcal{M}$ be an m-tuple of elements of F_{tree}. For any $\alpha \in F_{\text{tree}}$ define function $\hat{\beta}(\alpha) : \Omega_{\text{tree}} \to \mathcal{M}_n$ by*

$$\hat{\beta}(\alpha)((T,\omega)) = \begin{cases} \beta_{\alpha((T,\omega))}((T,\omega)) & \textit{if } \alpha((T,\omega)) < m \\ 0 & \textit{otherwise.} \end{cases}$$

Then this function is in F_{tree} too.

Proof: The proof is analogous to the proof of Lemma 7.2.2. First note that because the tuple of β_is is in $\mathcal{M}$ there must be a common upper bound $k \in \mathbf{N}$ on their level. Hence we may assume without loss of generality that α and all β_i are of the same level (the level can always be artificially increased by extending the trees Z in all possible ways).

Then note that the common parts $Z \in \text{Tree}_k$ of the trees computing random variables α and β_is on $\{T\} \times \{0,1\}^n$, $Z \preceq T$, can be shared. That is, if we use a tree S_Z to compute that the value of α on a sample is i and then we want to further compute the value of β_i at the sample, we need not go through the Z again. We just use the part of tree computing β_i below Z (it has a depth n raised to an infinitesimal). □

The lemma allows us to define family G_{tree} in the same way as G_{rud} was defined from F_{rud}.

Definition 10.2.6 Let $m \in \mathcal{M}_n$ and let $\hat{\beta} = (\beta_0, \ldots, \beta_{m-1}) \in \mathcal{M}$ be an m-tuple of elements of F_{tree}.

For any $\alpha \in F_{\text{tree}}$ define $\hat{\beta}(\alpha) : \Omega_{\text{tree}} \to \mathcal{M}_n$ by

$$\hat{\beta}(\alpha)((T,\omega)) = \begin{cases} \beta_{\alpha((T,\omega))}((T,\omega)) & \text{if } \alpha((T,\omega)) < m \\ 0 & \text{otherwise.} \end{cases}$$

Family G_{tree} consists of all random variables Θ computed by some $\hat{\beta} = (\beta_0, \ldots, \beta_{m-1})$, where $m \in \mathcal{M}_n$ and $\hat{\beta} \in \mathcal{M}$ is an m-tuple of elements of F_{tree}.

We conclude this section with a technical lemma.

Lemma 10.2.7 **(1)** *Let $A(x)$ be an open $L_n^2(F_{\text{tree}}, G_{\text{tree}})$-formula with x the only free variable. Then there are $k \in \mathbf{N}$ and $t > \mathbf{N}$ such that for all $i \in \mathcal{M}_n$ the membership in $\langle\langle A(i) \rangle\rangle$ is computed by a 0/1-valued element of F_{tree} having level k and the depth bounded above by $d_k + n^{1/t}$.*

(2) *Let $s > \mathbf{N}$ and $w := n^{1/s}$. Assume $(\alpha_1, \ldots, \alpha_w) \in \mathcal{M}$ is a w-tuple of elements of F_{tree}. Then there is $\beta \in F_{\text{tree}}$ computing simultaneously all α_i in the tuple (i.e. it computes the w-tuple of the values).*

(3) *Families F_{tree} and G_{tree} are closed under definition by cases by open $L_n^2(F_{\text{tree}}, G_{\text{tree}})$-formulas.*

Proof: Part (1) is proved, via Lemma 10.2.5, analogously to Lemma 8.2.1. For Part (2) we may assume, as in the proof of Lemma 10.2.5, that the level of all α_i is the same, some $k \in \mathbf{N}$.

Each of the w functions α_i is computed by a tuple of trees $(S_Z^i)_Z$, $Z \in \text{Tree}_k$, with the properties as in Definition 10.2.4. As the tuple is an element of $\mathcal{M}$ there is one $t > \mathbf{N}$ such the the depth of all trees, for any Z and i, is bounded above by $d_k + n^{1/t}$.

For any one $Z \in \text{Tree}_k$ one can compute the values of all α_i on the subset of the sample space $\bigcup_{T:Z \preceq T} \{T\} \times \{0,1\}^n \subseteq \Omega_{\text{tree}}$ by a tree S_Z constructed as follows: first compute as in Z and then successively as in the subtrees of S_Z^i starting at level d_k, for $i = 1, \ldots, w$. The depth of this tree is bounded above by $d_k + w \cdot n^{1/t}$. The term $w \cdot n^{1/t}$ is bounded by n raised to an infinitesimal, hence S_Z obeys the condition required in Definition 10.2.4.

Part (3) is now obvious. □

11

Quantifier elimination in $K(F_{\text{tree}}, G_{\text{tree}})$

Our first aim in this chapter is to show that the structure $K(F_{\text{tree}}, G_{\text{tree}})$ is Skolem closed and hence admits bounded quantifier elimination. We then observe that the argument yielding open comprehension and open induction in $K(F_{\text{rud}}, G_{\text{rud}})$ works here too and hence we will get comprehension and induction for $\Sigma_0^{1,b}$-formulas, and hence theory V_1^0, valid in $K(F_{\text{tree}}, G_{\text{tree}})$ by Theorem 9.3.1.

11.1 Skolem functions

Theorem 11.1.1 *The structure $K(F_{\text{tree}}, G_{\text{tree}})$ is Skolem closed.*

Proof: We need to show that for any open first-order $L_n^2(F_{\text{tree}}, G_{\text{tree}})$-formula $B(\overline{x}, y)$ and for any $m \in \mathcal{M}_n$ there is a $\Theta \in G_{\text{tree}}$ that is a Skolem function for $\exists y B(\overline{x}, y)$ on m.

Assume without loss of generality (we have a pairing function) that $\overline{x}$ is just one variable x. Assume we have an m-tuple $\hat{\beta} = (\beta_0, \ldots, \beta_{m-1}) \in \mathcal{M}$ of elements of F_{tree} such that for all $i < m$:

$$[\![\beta_i < m \wedge B(i, \beta_i)]\!] = [\![\exists y < m B(i, y)]\!].$$

Then we would like to take for the Skolem function the $\Theta \in G_{\text{tree}}$ computed by $\hat{\beta}$. Of course, $\Theta(i) = \beta_i$ and so Θ behaves as a Skolem function for $x = 0, \ldots, m-1$ but that is not enough. For example, it may happen that there is another m-tuple $\hat{\gamma} = (\gamma_0, \ldots, \gamma_{m-1}) \in \mathcal{M}$ of elements of F_{tree} behaving as $\hat{\beta}$ for all $i < m$:

$$[\![\beta_i < m \wedge B(i, \beta_i)]\!] = [\![\gamma_i < m \wedge B(i, \gamma_i)]\!]$$

but

$$\langle\langle \gamma_i < m \wedge B(i, \gamma_i)\rangle\rangle \setminus \langle\langle \beta_i < m \wedge B(i, \beta_i)\rangle\rangle$$

are non-empty. The only information we have about such difference sets is that their individual counting measures in Ω_{tree} are infinitesimal. But it may happen that there is an $\alpha \in F_{\text{tree}}$ with $[\![\alpha < m]\!] = 1_{\mathcal{B}}$ defining a distribution on $i < m$ so that the errors of individual β_is combine into a set with standard positive probability. That is, with a positive probability

$$\alpha(\omega) = i \wedge i \notin \langle\langle \beta_i < m \wedge B(i, \beta_i) \rangle\rangle.$$

By a careful choice of $\hat{\beta}$ we need to exclude this possibility.

For a $\hat{\beta}$ of the form as above define:

$$E_i := \langle\langle \exists y < m B(i,y) \rangle\rangle \setminus \langle\langle \beta_i < m \wedge B(i, \beta_i) \rangle\rangle.$$

Claim 1: *Assume that the counting measure (in Ω_{tree}) of $\bigcup_{i<m} E_i \in \mathcal{A}$ is infinitesimal. Then $\Theta \in G_{\text{tree}}$ computed by $\hat{\beta}$ is a Skolem function for $\exists y < m B(x,y)$ on m.*

Assume for the sake of contradiction that for some $\alpha \in F_{\text{tree}}$

$$[\![\Theta(\alpha) < m \wedge B(\alpha, \Theta(\alpha))]\!] < [\![\exists y < m B(\alpha, y)]\!].$$

Applying part (3) of Lemma 10.2.7 and Lemma 3.3.2 we get $\gamma \in F_{\text{tree}}$ such that

$$[\![\Theta(\alpha) < m \wedge B(\alpha, \Theta(\alpha))]\!] < [\![\gamma < m \wedge B(\alpha, \gamma)]\!].$$

Therefore

$$U := \langle\langle \gamma < m \wedge B(\alpha, \gamma) \rangle\rangle \setminus \langle\langle \Theta(\alpha) < m \wedge B(\alpha, \Theta(\alpha)) \rangle\rangle$$

has a non-infinitesimal counting measure in Ω_{tree}. But clearly

$$U \subseteq \bigcup_{i<m} \left(\{(T,\omega) \in \Omega_{\text{tree}} \mid \alpha((T,\omega)) = i\} \cap E_i \right) \subseteq \bigcup_{i<m} E_i,$$

which is a contradiction. This proves the claim.

So we want to find $\hat{\beta}$ such that the error sets E_i satisfy the hypothesis of Claim 1. For this we need to understand how the membership in $\langle\langle \exists y < m B(i,y) \rangle\rangle$ is computed. We have:

$$\langle\langle \exists y < m B(i,y) \rangle\rangle = \bigcup_{j<m} \langle\langle B(i,j) \rangle\rangle.$$

By Part (1) of Lemma 10.2.7 the membership in sets $\langle\langle B(i,j)\rangle\rangle$ is computed by 0/1-valued functions $\xi_{i,j}$ from F_{tree} such that all $\xi_{i,j}$ have the same standard level k and depth bounded above by $d_k + n^{1/t}$, for some common non-standard t.

Claim 2: *For any $i < m$, $0 \le u < v < m$ and $s > \mathbf{N}$ there are elements $\kappa_{i,u,v} \in F_{\text{tree}}$ of level $k+1$ and depth $d_{k+1} + n^{\frac{t+s}{ts}}$, such that:*

$$\text{Prob}_{(T,\omega)\in\Omega_{\text{tree}}}\left[((T,\omega) \in \bigcup_{u\le j\le v} \langle\langle B(i,j)\rangle\rangle) \not\equiv (\kappa_{i,u,v}((T,\omega)) = 1)\right]$$

$$< \left(\frac{10n^{1/t}(n-d_{k+1})}{n-d_k}\right)^{n^{1/s}} < \left(\frac{10n^{1/t}n^{\frac{1+2^{k+2}}{2^{k+3}}}}{n^{\frac{1+2^{k+1}}{2^{k+2}}}}\right)^{n^{1/s}} < 2^{-n^{1/s}}.$$

We shall use Theorem 10.1.2 to prove this key claim. Fix parameters i,u,v,s satisfying the restrictions of the claim. As i,u,v are fixed we ease the notation and do not reflect these parameters in the names of the forthcoming objects (trees and formulas). Also put $a := n^{1/t}$and $b := n^{1/s}$.

Let $(S^j_Z)_Z$, for $u \le j \le v$ and $Z \in \text{Tree}_k$, be the trees computing $\xi_{i,j}$. Hence each S^j_Z extends Z by attaching some trees of depth at most a to its leafs. Let $\mu^j_{Z,\ell}$ be the tree of depth at most a attached to a leaf ℓ of Z in S^j_Z, and let $\psi^j_{Z,\ell}$ be the disjunction of all terms corresponding to paths in $\mu^j_{Z,\ell}$ ending with label 1 (for yes). Note that each $\psi^j_{Z,\ell}$ is an a-DNF formula.

The element $\kappa_{i,u,v} \in F_{\text{tree}}$ will be computed by trees $(R_W)_{W\in\text{Tree}_{k+1}}$. Tree R_W will extend the $(k+1)$-tree W by attaching certain trees $\tau_{W,\ell'}$ to its leafs ℓ', each $\tau_{W,\ell'}$ of the depth at most $a \cdot b = n^{\frac{t+s}{ts}}$.

We are now going to argue, using Theorem 10.1.2, that there exist trees $\tau_{W,\ell'}$ such that the element $\kappa_{i,u,v}$ defined by the resulting trees $(R_W)_W$ satisfies the requirement of the claim.

Fix $Z \in \text{Tree}_k$ and its leaf ℓ. Let X_ℓ be the support of ℓ in Z, so $|X_\ell| = d_k$. Consider propositional formula

$$\varphi_{Z,\ell} := \bigvee_{u\le j\le v} \psi^j_{Z,\ell}.$$

It is an a-DNF formula, as $\psi^j_{Z,\ell}$ are. It has $n - d_k$ variables corresponding to elements of $[n] \setminus X_\ell$. By its definition the formula has the following property. Let $\Gamma_{Z,\ell} \subseteq \Omega_{\text{tree}}$ be the set of all $(T,\omega) \in \Omega_{\text{tree}}$ such that $Z \preceq T$ and ω leads in

Z to the leaf ℓ. Then for $(T,\omega) \in \Gamma_{Z,\ell}$:

$$\left((T,\omega) \in \bigcup_{u \le j \le v} \langle\langle B(i,j) \rangle\rangle \right) \text{ iff } (\omega \text{ satisfies } \varphi_{Z,\ell}).$$

For leafs ℓ of Z and sets $Y_\ell \subseteq [n] \setminus X_\ell$ of size $(n-d_k)^{\delta_{k+1}}$, all possible truth assignments on Y_ℓ are in a bijective correspondence with leafs ℓ' of a $(k+1)$-tree W extending Z by querying elements of Y_ℓ.

By Theorem 10.1.2 if we take a random such set $Y_\ell \subseteq [n] \setminus X_\ell$ and a random truth assignment $\sigma_{\ell'}$ (corresponding to leaf ℓ' in W) on Y_ℓ then we can find a decision tree $\tau_{W,\ell'}$ that

- is equivalent to $\varphi_{Z,\ell}$ for all ω that lead to ℓ' in W, and
- has depth at most $a \cdot b$

with the probability of failing to do so at most

$$\left(\frac{10a(n-d_{k+1})}{n-d_k} \right)^b$$

(the quantity required in the claim). The claim now follows by noting that:

1. sets $\Gamma_{Z,\ell}$, with $Z \in \text{Tree}_k$ and ℓ a leaf in Z, partition the sample space Ω_{tree}, and
2. sets $\Gamma_{W,\ell'}$, with $W \in \text{Tree}_{k+1}$ such that $Z \preceq W$ and ℓ' a leaf in W below ℓ, partition sets $\Gamma_{Z,\ell}$.

By the estimate above, suitable trees $\tau_{W,\ell'}$ exist for all but a fraction of at most $(10a(n-d_{k+1})/(n-d_k))^b$ pairs (W,ℓ') considered in part 2, and hence by 1 this also estimates the fraction of all pairs (W,ℓ'), $W \in \text{Tree}_{k+1}$ and leaf ℓ' in W for which this happens. This proves the claim.

To define elements β_i that produce a j such that $(T,\omega) \in \langle\langle B(i,j) \rangle\rangle$ if j exists (with only a small probability of failing) we simulate the binary search, computing functions $\kappa_{i,u,v}$. This requires us to compute $\log(m)$ of such functions on a path in the search, and the whole binary search can be performed by an element β_i of F_{tree} (as in Lemma 10.2.7, Part 2). The depth of the trees computing β_i is bounded above by $d_{k+1} + \log(m) \cdot a \cdot b$ and that obeys the restrictions for a $(k+1)$-level element of F_{tree}.

By Claim 2 each $\kappa_{i,u,v}$ makes an error with probability less than $2^{-n^{1/s}}$. There is at most m such pairs $u \le v$ used in a binary search and there are m different instances i. Hence the counting measure of the error set $\bigcup_{i<m} E_i$ for

$\hat{\beta} = (\beta_0, \ldots, \beta_{m-1})$ defined in this way is less than

$$2^{-n^{1/s}} \cdot m^2.$$

By taking s non-standard but small enough this quantity can be made infinitesimal. □

The theorem yields, via Theorem 9.2.3, the bounded quantifier elimination.

Corollary 11.1.2 *Model $K(F_{\text{tree}}, G_{\text{tree}})$ admits bounded quantifier elimination.*

We shall note for future reference one more corollary of the construction.

Corollary 11.1.3 *For any Σ^b_∞-formula $A(x_1, \ldots, x_k)$, possibly with parameters from F_{tree} or G_{tree}, and for any $m \in \mathcal{M}_n$ and arbitrarily small but non-standard $s > \mathbf{N}$ there is an open $L^2_n(F_{\text{tree}}, G_{\text{tree}})$-formula $B(x_1, \ldots, x_k)$ such that*

$$[\![\forall x_1, \ldots, x_k < m\ (A(x_1, \ldots, x_k) \equiv B(x_1, \ldots, x_k))]\!] = 1_{\mathcal{B}}$$

and such that for any $\alpha_1, \ldots, \alpha_k \in F_{\text{tree}}$ the counting measure of the symmetric difference

$$\left\langle\!\left\langle \bigwedge_i \alpha_i < m \to A(\alpha_1, \ldots, \alpha_k) \right\rangle\!\right\rangle \triangle \left\langle\!\left\langle \bigwedge_i \alpha_i < m \to B(\alpha_1, \ldots, \alpha_k) \right\rangle\!\right\rangle$$

is less than $2^{-n^{1/s}}$. In particular, it is infinitesimal.

11.2 Comprehension and induction for $\Sigma^{1,b}_0$-formulas

Lemma 11.2.1 *Open comprehension and open induction are valid in $K(F_{\text{tree}}, G_{\text{tree}})$.*

Proof: The proof is identical to the proof of Corollary 8.2.3 and Lemma 8.3.2 (via Lemma 8.3.1), with Lemma 10.2.7 replacing Lemma 8.2.1, and noting (again using Lemma 10.2.7) that the witness α constructed as in the proof of Lemma 8.3.1 is in F_{tree}. □

Lemma 11.2.1 and Corollary 11.1.2 imply via Theorem 9.3.1 the following statement.

Theorem 11.2.2 *Comprehension and induction for $L^2_n(F_{\text{tree}}, G_{\text{tree}})$ $\Sigma^{1,b}_0$-formulas is valid in $K(F_{\text{tree}}, G_{\text{tree}})$. In particular, theory V^0_1 is valid in the structure.*

12

Witnessing, independence and definability in V_1^0

A proof-theoretic approach to witnessing theorems for V_1^0 is to Skolemize the theory first via AC^0 maps on strings and then use some form of Herbrand theorem (see also the remark at the end of Chapter 6). In this chapter we shall prove a witnessing theorem for V_1^0 by applying the general witnessing theorems from Chapter 3 to model $K(F_{\mathrm{tree}}, G_{\mathrm{tree}})$. The result corresponds to the one obtained by the classical approach combined with the switching lemma (Theorem 10.1.2).

Two prominent types of formula $A(x)$ to consider witnessing theorems for are:

(a) $\forall X < x \exists Y < x \Sigma_0^{1,b}$,
(b) $\forall X < x \exists Y < x \forall Z < x \Sigma_0^{1,b}$.

We shall prove a witnessing theorem for the first type of formula for V_1^0 in this chapter, and for the second type of formula for theory $Q_2 V_1^0$ in Chapter 16. It will be clear that the proofs work identically for both theories, and both witnessing theorems apply equally well to V_1^0 and $Q_2 V_1^0$. We use the witnessing theorems to give true statements of each form unprovable in the respective theory. More examples of witnessing theorems for stronger theories (PV, S_2^1, and T_2^1) are given in Part VII. All these examples show, we believe, that forcing with random variables provides an alternative framework for proving witnessing theorems.

There are three other types of formula $A(x)$,

(c) $\forall X < x \Sigma_0^{1,b}$ (i.e. $s\Pi_1^{1,b}$),
(d) $\forall X < x \exists y < x \forall Z < x \Sigma_0^{1,b}$,
(e) $\exists Y < x \forall Z < x \Sigma_0^{1,b}$,

that are very interesting from the point of view of proof complexity. But independence for sentences of this form cannot be proved using $K(F_{\mathrm{tree}}, G_{\mathrm{tree}})$ because, as we shall show in Section 12.2, the model preserves all true $s\Pi_1^{1,b}$-sentences. To avoid such a preservation property is a key issue if one hopes to use models of bounded arithmetic for lengths-of-proofs lower bounds. We shall return to this topic in Part V.

In Section 12.3 we note that the $s\Pi_1^{1,b}$-elementarity, while bad for proof complexity, allows us to prove a circuit lower bound for AC^0 circuits computing parity. This argument corresponds to the classical proof via the switching lemma.

12.1 Witnessing $\forall X < x \exists Y < x \Sigma_0^{1,b}$-formulas

Let $A(x)$ be a $\forall X < x \exists Y < x \Sigma_0^{1,b}$-formula of the form

$$\forall X < x \exists Y < x B(x, X, Y)$$

with B a $\Sigma_0^{1,b}$-formula. Families F_{tree} and G_{tree} are closed under definition by open formulas (Lemma 10.2.7) and so we may apply Lemma 3.3.3 to deduce the following statement for $K(F_{\text{tree}}, G_{\text{tree}})$.

Lemma 12.1.1 *Assume* $[\![A(n)]\!] = 1_{\mathcal{B}}$. *Then for every* $\Delta \in G_{\text{tree}}$ *and every standard* $\epsilon > 0$ *there is* $\Gamma \in G_{\text{tree}}$ *such that*

$$\mu([\![\Delta < n \rightarrow (\Gamma < n \wedge B(n, \Delta, \Gamma))]\!]) > 1 - \epsilon.$$

Applying Corollary 11.1.2 to $B(n, \Delta, \Gamma)$ and then Lemma 2.2.1 we obtain the next statement.

Lemma 12.1.2 *Assume* $[\![A(n)]\!] = 1_{\mathcal{B}}$. *Then for every* $\Delta \in G_{\text{tree}}$ *and every standard* $\epsilon > 0$ *there is* $\Gamma \in G_{\text{tree}}$ *such that*

$$\text{Prob}_{(T,\omega) \in \Omega_{\text{tree}}}[\, (T, \omega) \in \langle\!\langle \Delta < n \rightarrow (\Gamma < n \wedge B(n, \Delta, \Gamma)) \rangle\!\rangle \,] > 1 - \epsilon.$$

Applying the lemma to a specific random variable Δ will give us the witnessing theorem. Δ is simply the projection on the second coordinate:

$$\Delta((T, \omega)) := \omega.$$

If the samples were just ωs and not pairs (T, ω) this would be simply the identity map (which represents in a sense the uniform distribution on the sample space). It is easy to see that indeed $\Delta \in G_{\text{tree}}$; it is given by a tuple $\hat{\beta} = (\beta_i)_{i<n}$ where β_i

is a tree querying $\omega_i =_? 0$ and labeling the two resulting leafs by the respective answers 0 or 1 to the query. Also note that $\langle\langle \Delta < n \rangle\rangle = \Omega_{\text{tree}}$.

Theorem 12.1.3 *Assume $[[A(n)]] = 1_{\mathcal{B}}$. Then for every standard $\epsilon > 0$ there is $\Gamma \in G_{\text{tree}}$ such that*

$$\text{Prob}_{(T,\omega)\in\Omega_{\text{tree}}}[\ \Gamma((T,\omega)) < n \wedge B(n,\omega,\Gamma((T,\omega)))\ ] \ > \ 1-\epsilon.$$

In particular, this holds if theory V_1^0 proves $\forall x A(x)$.

Using compactness, Lemma 10.2.2 and the fact that in the definition of cut $\mathcal{M}_n$ and of subsequent structures we have not used any property of n except that it is non-standard, we may formulate a finitary corollary of Theorem 12.1.3.

Corollary 12.1.4 *Assume that theory V_1^0 proves $\forall x A(x)$. Then for every standard $\epsilon > 0$ there is a collection $W = (W_i)_{i<m}$ of trees W_i of depth less than $m - m^{1/2}$ such that for all $m \in \mathbf{N}$ large enough:*

$$\text{Prob}_{\omega\in\{0,1\}^m}[\ B(m,\omega,W(\omega))\ ] \ > \ 1-\epsilon.$$

A specific formula $A(x)$ for which the witnessing theorem yields an unprovability from V_1^0 (and which is interesting in connection with theory $Q_2V_1^0$ defined in Chapter 13) expresses the existence of a set Y counting parity of the given set X (in a slightly more general situation with a family of sets rather than one set, this will be called a 2-cover in Section 13.2). The formula is defined by taking for $B(x,X,Y)$ the formula to be denoted $\text{Par}(x,X,Y)$:

$$0 \notin Y \wedge \forall y < x (y+1 \in Y \equiv (y \in X \oplus y \in Y))$$

where $C \oplus D$ abbreviates $C \not\equiv D$. The intended meaning of the formula is that $y \in Y$ is true iff the number of $z < y$ such that $z \in X$ is odd.

It is now easy to see that either Theorem 12.1.3 or directly Corollary 12.1.4 entail the following unprovability result.

Theorem 12.1.5 *Sentence $\forall x \forall X < x \exists Y < x \text{Par}(x,X,Y)$ is true but unprovable in theory V_1^0.*

12.2 Preservation of true $s\Pi_1^{1,b}$-sentences

We are now going to show that true sentences $\forall x A(x)$ for formulas $A(x)$ of one of the forms (c), (d), or (e) listed in the introduction to Chapter 12 are valid in model $K(F_{\text{tree}}, G_{\text{tree}})$. The key case is (c) the other two are its corollaries.

Theorem 12.2.1 *Assume sentence $\forall x A(x)$ is true where $A(x)$ is a $s\Pi_1^{1,b}$-formula. Then $\forall x A(x)$ is valid in $K(F_{\text{tree}}, G_{\text{tree}})$.*

Proof: Assume $A(x)$ is of the form $\forall X < x B(x, X)$ where B is a $\Sigma_0^{1,b}$ formula. It suffices to show that

$$[\![\forall x < m B(x, \Gamma)]\!] = 1_{\mathcal{B}}$$

for any $m \in \mathcal{M}_n$ and $\Gamma \in G_{\text{tree}}$.

By Corollary 11.1.3 there is an open $L_n^2(F_{\text{tree}}, G_{\text{tree}})$-formula $C(x)$ such that

$$[\![\forall x < m\ (B(x, \Gamma) \equiv C(x))]\!] = 1_{\mathcal{B}}$$

and the counting measure of the symmetric difference

$$\langle\!\langle \alpha < m \to B(\alpha, \Gamma) \rangle\!\rangle \,\triangle\, \langle\!\langle \alpha < m \to C(\alpha) \rangle\!\rangle$$

is infinitesimal for any $\alpha \in F_{\text{tree}}$.

Because $\forall x A(x)$ is true, $\langle\!\langle \alpha < m \to B(\alpha, \Gamma) \rangle\!\rangle = \Omega_{\text{tree}}$. As C is open we may deduce that $[\![\alpha < m \to C(\alpha)]\!] = 1_{\mathcal{B}}$, for any α. This proves the theorem. □

We shall use the theorem to deduce the preservation for formulas of types (d) and (e) too. We state it in two separate corollaries as their proofs are different.

Corollary 12.2.2 *Assume sentence $\forall x A(x)$ is true where $A(x)$ is a formula of the form $\forall X < x \exists y < x \forall Z < x \Sigma_0^{1,b}$.*

Then $\forall x A(x)$ is valid in $K(F_{\text{tree}}, G_{\text{tree}})$.

Proof: Let $A(x)$ be a formula of the form $\forall X < x \exists y < x \forall Z < x C(x, X, y, Z)$ where C is a $\Sigma_0^{1,b}$ formula. If $\forall x A(x)$ is true so is the formula $\forall x A^*(x)$ where $A^*(x)$ is

$$\forall X < x \forall Z^* < x^2 \exists y < x\ C^*(x, X, y, Z^*)$$

where we think of Z^* as an array of sets $Z_y^* < x$, one for each $y < x$, and $C^*(x, X, y, Z^*)$ says that $C(x, X, y, Z_y^*)$ holds.

This is a formula of the form to which Theorem 12.2.1 applies, so $\forall x A^*(x)$ is valid in $K(F_{\text{tree}}, G_{\text{tree}})$. The validity of $\forall x A(x)$ follows, noting that the bounded collection scheme is valid in the model. We state only the claim we need.

Claim: *The sentence*

$$\forall x, X\ [\ (\forall y < x \exists Z < x \neg C(x, X, y, Z)) \to (\exists Z^* < x^2 \forall y < x \neg C^*(x, X, y, Z^*))\ ]$$

is valid in $K(F_{\text{tree}}, G_{\text{tree}})$.

The claim is proved using the collection scheme in $\mathcal{M}$. □

Corollary 12.2.3 *Assume sentence $\forall x A(x)$ is true where $A(x)$ is a formula of the form $\exists Y < x \forall Z < x \Sigma_0^{1,b}$.*

Then $\forall x A(x)$ is valid in $K(F_{\text{tree}}, G_{\text{tree}})$.

Proof: Let $A(x)$ be a formula of the form $\exists Y < x \forall Z < x\ D(x, Y, Z)$ where D is a $\Sigma_0^{1,b}$-formula. If $\forall x A(x)$ is true then there exists a binary relation R on $\mathbf{N}$ (i.e. we have a symbol for it in L_n) such that for each $u \in \mathcal{M}$ the set R_u witnesses the truth of $\exists Y < u \forall Z D(u, Y, Z)$. Here we again think of R as encoding an (infinite, this time) array of sets. In other words, sentence

$$\forall x (R_x < x \wedge \forall Z < x D(x, R_x, Z))$$

is true. But this sentence is $s\Pi_0^{1,b}$ and hence by Theorem 12.2.1 it is valid in $K(F_{\text{tree}}, G_{\text{tree}})$.

Clearly, the validity of this sentence implies (in predicate logic only) the validity of $\forall x A(x)$. □

12.3 Circuit lower bound for parity

In Boolean complexity theory one of the well-known results states an exponential lower bound on constant-depth circuits computing parity. It was proved gradually by Ajtai [1], Furst *et al.* [34], Yao [108] and Hastad [40].

Theorem 12.3.1 *For every $d \geq 1$ there is an $\epsilon > 0$ such that, for $k >> 0$, any depth d circuit computing parity of k bits must have the size at least 2^{k^ϵ}.*

We shall observe in this section that the lower bound can be proved by using the structure $K(F_{\text{tree}}, G_{\text{tree}})$. Of course, this is not surprising because the heart of Theorem 12.3.1 is the switching lemma (Theorem 10.1.2) which also plays a key role in the analysis of $K(F_{\text{tree}}, G_{\text{tree}})$. However, it is interesting to see that the $s\Pi_1^{1,b}$-elementarity of $K(F_{\text{tree}}, G_{\text{tree}})$ over the standard model, which renders the structure useless for lengths-of-proofs lower bounds, allows us on the other hand to prove a circuit lower bound.

Basic facts from descriptive complexity theory imply that a set $\mathcal{U}$ of strings (i.e. of bounded sets) is definable by constant-depth circuits of a subexponential size iff there is an L_n $\Sigma_0^{1,b}$-formula $A(x,X)$ (with no parameters) such that for all $m \in \mathbf{N}$ and $V \subseteq [m]$ the following holds:

$$V \in \mathcal{U} \cap \{0,1\}^m \text{ iff } A(m,V) \textit{ is true.}$$

Take for $\mathcal{U}$ the set of sets of odd cardinality and assume for the sake of contradiction that there is a formula $A(x,X)$ as above. If that is so then

$$\forall x \forall X < x \forall y < x\, [A(x,X) \not\equiv A(x, X \oplus \{y\})]$$

holds, where $X \oplus \{y\}$ is an abbreviation for the set resulting from adding y to X if $y \notin X$ or removing it if $y \in X$. This is a $s\Pi_1^{1,b}$-sentence and thus by Theorem 12.2.1 we may conclude that it is also valid in $K(F_{\text{tree}}, G_{\text{tree}})$. We bring the assumption to a contradiction by showing that, in fact, for some $\alpha \in F_{\text{tree}}$

$$(*) \qquad [\![\alpha < n \wedge (A(n,\Delta) \equiv A(n, \Delta \oplus \{\alpha\}))]\!] \;=\; 1_{\mathcal{B}}.$$

Here $\Delta \in G_{\text{tree}}$ is the random variable considered in Theorem 12.2.1, the projection on the second coordinate.

By Corollary 11.1.3 there is $\gamma \in F_{\text{tree}}$ such that $[\![\gamma = 1]\!] = [\![A(n,\Delta)]\!]$. Using this element define α by extending the tree S defining γ as follows. Let ℓ be a leaf of S and let $i_\ell \in [n]$ be the first element i of $[n]$ such that $\omega_i =_? 0$ is not queried on the path to ℓ. Such an i exists as the depth of S is less than n. The tree computing α is the tree S with all ℓ labeled by i_ℓ.

It is not difficult to see that with this α the equality $(*)$ holds.

PART IV

$AC^0(2)$ world

13
Theory $Q_2V_1^0$

13.1 Q_2 quantifier and theory $Q_2V_1^0$

Modular counting quantifiers were considered first in bounded arithmetic by Paris and Wilkie [86]. We shall consider just one of them, the parity quantifier.

The new quantifier will be denoted Q_2. It can be used only as a bounded quantifier: if A is a formula then we can form a new formula $Q_2y < t(\overline{x})A$. The intended meaning is: $Q_2y < t(\overline{x})A$ is true iff there is an odd number of $y < t(\overline{x})$ for which A is true.

The axioms governing the behavior of Q_2 are:

1. $\neg Q_2y < 0A(y)$,
2. $(Q_2y < tA(y) \wedge A(t)) \rightarrow \neg\, Q_2y < (t+1)A(y)$,
3. $(Q_2y < tA(y) \wedge \neg A(t)) \rightarrow Q_2y < (t+1)A(y)$.

We shall denote by $Q_2\Sigma_0^{1,b}$, $Q_2\Sigma_\infty^b$ and similar the corresponding classes of bounded formulas allowing also the Q_2 quantifier. In particular, $Q_2\Sigma_0^{1,b}$ is the class of bounded L^2-formulas without second-order quantifiers while $Q_2\Sigma_\infty^b$ is the class of first-order bounded formulas.

Theory $Q_2V_1^0$ is defined quite analogously to V_1^0. The axioms of $Q_2V_1^0$ are:

- axioms of PA^- from Chapter 6
- axioms 1–3 above for the Q_2-quantifier
- comprehension axioms for all $Q_2\Sigma_0^{1,b}$-formulas
- induction axioms for all $Q_2\Sigma_0^{1,b}$-formulas.

Note that it is well known (as a consequence of Theorem 12.3.1) that the Q_2-quantifier cannot be defined by a $\Sigma_0^{1,b}$-formula, even with extra parameters.

13.2 Interpreting Q_2 in structures

The truth values of sentences $\exists y A(y)$ and $\forall y A(y)$ are determined by the truth values of all instances $A(\alpha)$, computing in $\mathcal{B}$ the supremum and the infimum of the set of values, respectively. However, it is difficult to see how one could define $[\![Q_2 y < \beta A(y)]\!]$ coherently from values of its instances $[\![\alpha < \beta \wedge A(\alpha)]\!]$, even if all these values were equal to $1_{\mathcal{B}}$. For example, the family F_{alg} (see the next chapter) from which α may be drawn is infinite and there is a bijective correspondence between the family and the family minus one element.

In interpreting the Q_2 quantifier we take an indirect approach. We notice that the quantifier can be defined by a $\Sigma_0^{1,b}$-formula (in a sense to be described shortly) as long as the class of relations of the structure is rich enough. We will illustrate this with the following informal example (a formal definition for the specific case of $K(F_{\text{alg}}, G_{\text{alg}})$ will appear in Section 15.2).

Assume $R(x_1, \ldots, x_k, y) \in \mathcal{M}$ is a relation on $m \in \mathcal{M}_n$. We shall call any relation $S(x_1, \ldots, x_k, y) \in \mathcal{M}$ on m a **2-cover** of R iff the following holds: For $u_1, \ldots, u_k < m$ and $v < m$

$$S(\overline{u}, v) \text{ iff } |\{w < v \mid R(\overline{u}, w)\}| \textit{ is odd.}$$

This makes a good sense in $\mathcal{M}$ where we can count the size of finite sets but in a general structure such a definition would not work. However, it is easy to see that the condition is equivalent to the validity of three other conditions, all first-order expressible:

1. $(\forall z < y \neg R(\overline{x}, z)) \rightarrow \neg S(\overline{x}, y)$,
2. $S(\overline{x}, y) \rightarrow (S(\overline{x}, y+1) \equiv \neg R(\overline{x}, y))$,
3. $\neg S(\overline{x}, y) \rightarrow (S(\overline{x}, y+1) \equiv R(\overline{x}, y))$.

Note that if we had the Q_2 quantifier we could define the 2-cover explicitly:

$$S(\overline{x}, y) \equiv Q_2 z < y R(\overline{x}, z).$$

The point is, however, that the same equivalence can be used to interpret the Q_2 quantifier in any structure (and in front of any $Q_2 \Sigma_0^{1,b}$-formula) as long as any $\Sigma_0^{1,b}$-definable relation R has a 2-cover on any m.

This construction is performed formally in Chapter 15. The 2-covers are defined there, essentially, by defining the parity of a set of Boolean values but only of an $\mathcal{M}$-finite set and in Boolean algebra $\mathcal{A}$ and not $\mathcal{B}$.

We note that in this way one can define a theory for the computational complexity class $AC^0(2)$ in the language of V^0 (i.e. without the Q_2 quantifier), having the same $\Sigma_0^{1,b}$-consequences as $Q_2 V_1^0$ does. This approach is taken in Cook and Nguyen [30].

14
Algebraic model

In this chapter we will construct a particular model $K(F_{alg}, G_{alg})$ based on algebraic decision trees.

14.1 Family F_{alg}

Recall that we have a fixed non-standard number n. Our sample space will be simple this time: $\Omega := \{0,1\}^n$.

Definition 14.1.1 Let $\mathbf{F}_2[x_1,\dots,x_n]$ be the ring of polynomials over $\mathbf{F}_2$ in $\mathcal{M}$. We shall denote by $\mathbf{F}_2^{\mathrm{low}}[x_1,\dots,x_n]$ the family of polynomials from $\mathbf{F}_2[x_1,\dots,x_n]$ of degree bounded above by some term of the form $n^{1/t}$, for a non-standard $t > \mathbf{N}$.

In other words, the degree of polynomials in $\mathbf{F}_2^{\mathrm{low}}[x_1,\dots,x_n]$ is subpolynomial in n. Family $\mathbf{F}_2^{\mathrm{low}}[x_1,\dots,x_n]$ is a subset of $\mathcal{M}$ but not a definable one.

To define the family F_{alg} forming the first-order part of the model we have to define (in $\mathcal{M}$) first an auxiliary concept of an algebraic decision tree.

Definition 14.1.2 An **algebraic decision tree** (or a decision tree for short) T over $\mathbf{F}_2^{\mathrm{low}}[x_1,\dots,x_n]$ is a finite binary tree whose inner nodes (i.e. non-leafs) are labeled by queries of the form $p(\overline{x}) =_? 0$, for $p(\overline{x}) \in \mathbf{F}_2^{\mathrm{low}}[x_1,\dots,x_n]$. The two outgoing edges from such a node are labeled yes and no, respectively.

The depth of a tree, denoted $\mathrm{dp}(T)$, is the length of the longest path from the root to a leaf.

The degree of the tree, denoted $\deg(T)$, is the maximum degree of a polynomial that is queried in T.

A labeling of T is a map $\ell \in \mathcal{M}$ assigning to leafs of T values in $\mathcal{M}_n$.

An algebraic tree is labeled a decision tree over $\mathbf{F}_2^{\mathrm{low}}[x_1,\dots,x_n]$.

Let $(T,\ell) \in \mathcal{M}$ be an algebraic tree over $\mathbf{F}_2^{\text{low}}[x_1,\ldots,x_n]$. Any $\omega \in \Omega$ induces an evaluation $x_i := \omega_i$ and thus determines a unique path through T ending in a leaf that we shall denote $T(\omega)$. In this way (T,ℓ) defines a function

$$\omega \in \Omega \rightarrow \ell(T(\omega)) \in \mathcal{M}_n.$$

Definition 14.1.3 The family F_{alg} consists of all functions

$$\alpha : \Omega \rightarrow \mathcal{M}_n$$

from $\mathcal{M}$ such that there is a labeled decision tree $(T,\ell) \in \mathcal{M}$ over ring $\mathbf{F}_2^{\text{low}}[x_1,\ldots,x_n]$ such that

- $\alpha(\omega) = \ell(T(\omega))$, for all $\omega \in \Omega$, and
- both the depth $\text{dp}(T)$ and the degree $\deg(T)$ of T are bounded above by a term of the form $n^{1/t}$, for some $t > \mathbf{N}$.

Note that a bound to $\deg(T)$ of the form $n^{1/t}$, $t > \mathbf{N}$, follows already from the fact that $(T,\ell) \in \mathcal{M}$ and that all query polynomials are in $\mathbf{F}_2^{\text{low}}[x_1,\ldots,x_n]$.

There is an alternative presentation of family F_{alg} that will be useful to keep in mind.

Lemma 14.1.4 *For every $\alpha \in F_{\text{alg}}$ there is a sequence of polynomials*

$$(p_0,\ldots,p_{u-1}) \in \mathcal{M}$$

from $\mathbf{F}_2^{\text{low}}[x_1,\ldots,x_n]$ (as the sequence is in $\mathcal{M}$ there is a common bound $n^{1/s}$, $s > \mathbf{N}$, for degrees of p_is) for some $u \in \mathcal{M}_n$ such that:

1. *Equations*
   (a) $\sum_{i<u} p_i = 1$
   (b) $p_i \cdot p_j = 0$*, for* $i \neq j \in u$
   hold identically on Ω.
2. *All labels of α are from $u = \{0,\ldots,u-1\}$.*
3. *For all $\omega \in \Omega$ and $i < u$:*

$$\alpha(\omega) = i \;\textit{ iff }\; p_i(\omega) = 1.$$

On the other hand, for all sequences $(p_0,\ldots,p_{u-1}) \in \mathcal{M}$ of polynomials from $\mathbf{F}_2^{\text{low}}[x_1,\ldots,x_n]$ of length $u \in \mathcal{M}_n$ obeying condition 1 there is $\alpha \in F_{\text{alg}}$ for which conditions 2 and 3 also hold.

Proof: Let (T,ℓ) be a tree defining α. For any leaf L in T define polynomial q_L as follows: label any yes-edge on the path to L leaving a query node $p =_? 0$ by $1-p$ and any no-edge by p, and take for q_L the product of the labels of the edges on the path. Note that $q_L \in \mathbf{F}_2^{\mathrm{low}}[x_1,\ldots,x_n]$ as its degree is bounded above by $\deg(T) \cdot \mathrm{dp}(T)$.

Clearly $q_L(\omega) = 1$ iff ω determines the path to L. Hence $\sum_L q_L$ is identically 1 on Ω and any two such polynomials attached to two different leafs are incompatible (i.e. at most one is non-zero).

Define p_i to be the sum of q_L for all leafs L labeled by i. This proves the first part of the lemma.

For the opposite direction, starting with the sequence $(p_0,\ldots,p_{u-1}) \in \mathcal{M}$, define α searching for a non-zero p_i by binary search, querying

$$\sum_{e \le i \le f} p_i =_? 0$$

for suitable $0 \le e < f < u$. The required bounds to degrees of polynomials and to the depth of the tree defining α are obvious as $u \in \mathcal{M}_n$. □

Given polynomials $(p_i)_{i \in u}$ satisfying equations $1a,b$ of Lemma 14.1.4 we shall denote the function α they define in the sense of the lemma as

$$\alpha := \sum_{i \in u} p_i \cdot i^*.$$

This is to be understood as a formal notation not as an actual sum in some algebraic expression. In particular, i^* is a symbol attached to an element $i \in \mathcal{M}_n$, not an element of any ring.

14.2 Family G_{alg}

The family G_{alg} is introduced quite analogously to the way families G_{rud} and G_{tree} were introduced earlier.

Definition 14.2.1 Let $m \in \mathcal{M}_n$ and let $\hat{\beta} = (\beta_0,\ldots,\beta_{m-1}) \in \mathcal{M}$ be an $\mathcal{M}$-tuple of elements of F_{alg}.

For any $\alpha \in F_{\mathrm{alg}}$ define $\hat{\beta}(\alpha) : \Omega \to \mathcal{M}_n$ by

$$\hat{\beta}(\alpha)(\omega) = \begin{cases} \beta_{\alpha(\omega)}(\omega) & \text{if } \alpha(\omega) < m \\ 0 & \text{otherwise.} \end{cases}$$

Lemma 14.2.2 *Let $m \in \mathcal{M}_n$ and let $\hat{\beta} = (\beta_0, \ldots, \beta_{m-1}) \in \mathcal{M}$ be an m-tuple of elements of F_{alg}.*

Then function $\hat{\beta}$ from Definition 14.2.1 maps F_{alg} into F_{alg}.

Proof: As $\hat{\beta} \in \mathcal{M}$, the induction in $\mathcal{M}$ implies that there is a maximal depth of a tree computing one of β_i, $i < m$. This depth is therefore bounded above by n raised to an infinitesimal. The depth of the tree computing any $\alpha \in F$ is bounded by a term of the same form, and hence so is the depth of the tree computing $\hat{\beta}(\alpha)$. □

Definition 14.2.3 Family G_{alg} consists of all random variables Θ computed by some $\hat{\beta} = (\beta_0, \ldots, \beta_{m-1})$, where $m \in \mathcal{M}_n$ and $\hat{\beta} \in \mathcal{M}$ is an m-tuple of elements of F_{alg}.

In other words, for Θ computed by $\hat{\beta}$ the function Θ_ω has the graph $\{(i, \beta_i(\omega)) \mid i < m\}$ (and is zero outside $[m]$).

Note that both families F_{alg} and G_{alg} are compact.

14.3 Open comprehension and open induction

The proof that open induction is valid in the structure $K(F_{\text{rud}}, G_{\text{rud}})$ applies literally here, by just replacing Boolean decision trees with algebraic trees. We just state the two relevant statements, for later reference.

The first one is analogous to Lemmas 8.2.1 and 10.2.7.

Lemma 14.3.1 **(1)** *Let $A(x)$ be an open $L_n^2(F_{\text{alg}}, G_{\text{alg}})$-formula with x the only free variable.*

Then there is $t > \mathbf{N}$ such that for all $i \in \mathcal{M}_n$ the membership in $\langle\langle A(i) \rangle\rangle$ is computed by an algebraic tree with the leafs labeled by 1 (for yes*) or 0 (for* no*) of the depth and the degree bounded above by $n^{1/t}$. In particular, the membership is computed by an element of F_{alg}.*

(2) *Let $s > \mathbf{N}$ and $w := n^{1/s}$. Assume $(\alpha_1, \ldots, \alpha_w) \in \mathcal{M}$ is an w-tuple of elements of F_{alg}. Then there is $\beta \in F_{\text{alg}}$ computing simultaneously all α_i in the tuple (i.e. it computes the w-tuple of the values).*

(3) *Families F_{alg} and G_{alg} are closed under definition by cases by open $L_n^2(F_{\text{alg}}, G_{\text{alg}})$-formulas.*

The next lemma is proved (using Lemma 14.3.1 in place of Lemma 8.2.1) in the same way as Lemmas 8.2.3 and 8.3.2.

Lemma 14.3.2 *Open comprehension and open induction are valid in $K(F_{\text{alg}}, G_{\text{alg}})$.*

15
Quantifier elimination and the interpretation of Q_2

In this chapter we show that any formula of the form

$$\mathbf{Q}y < t(\overline{x})A(\overline{x}, y),$$

where A is open and $\mathbf{Q}$ is one of quantifiers $\exists$, $\forall$ or Q_2, is equivalent in $K(F_{\text{alg}}, G_{\text{alg}})$ on any interval $[0, m)$ to an open formula (with free variables $\overline{x}$); that is, the equivalence is valid in $K(F_{\text{alg}}, G_{\text{alg}})$. For the Q_2 quantifier this is done by defining it suitably in $K(F_{\text{alg}}, G_{\text{alg}})$, and by verifying that its required properties become valid in $K(F_{\text{alg}}, G_{\text{alg}})$ under the interpretation. For the ordinary first-order quantifiers this requires a bit of work; the elimination of such quantifiers utilizes the Razborov–Smolensky approximation method from circuit complexity theory.

15.1 Skolemization and the Razborov–Smolensky method

Recall Definitions 9.2.1 and 9.2.2. As F_{alg} is closed under definition by cases by open $L_n(F_{\text{alg}}, G_{\text{alg}})$-formulas, and families F_{alg} and G_{alg} are compact, Theorem 3.5.2 guarantees the existence of $\gamma \in F_{\text{alg}}$ such that

$$[\![\exists y < mB(\alpha, y)]\!] = [\![\gamma < m \wedge B(\alpha, \gamma)]\!]$$

for any formula B. But, as discussed in Section 9.2, this is insufficient for a useful quantifier elimination, and we will proceed differently.

Theorem 15.1.1 *The structure $K(F_{\text{alg}}, G_{\text{alg}})$ is Skolem closed.*

Proof: The general strategy of the proof, as expressed in Claim 1, is identical to the proof of Theorem 11.1.1. For the sake of completeness of the presentation of the proof we repeat this common part here.

We need to show that for any open $L_n^2(F_{\text{alg}}, G_{\text{alg}})$-formula $B(x,y)$ (without loss of generality we assume that we have just one variable x) and for any $m \in \mathcal{M}_n$ there is a $\Theta \in G_{\text{alg}}$ that is a Skolem function for $\exists y B(\overline{x}, y)$ on m. Assume we have an m-tuple $\hat{\beta} = (\beta_0, \ldots, \beta_{m-1}) \in \mathcal{M}$ of elements of F_{alg} such that for all $i < m$

$$[\![\beta_i < m \wedge B(i, \beta_i)]\!] = [\![\exists y < m B(i,y)]\!].$$

Then we would like to take for the Skolem function the $\Theta \in G_{\text{alg}}$ computed by $\hat{\beta}$. Of course, $\Theta(i) = \beta_i$ and so Θ behaves as a Skolem function for $x = 0, \ldots, m-1$ but that is not enough. For example, it may happen that there is another m-tuple $\hat{\gamma} = (\gamma_0, \ldots, \gamma_{m-1}) \in \mathcal{M}$ of elements of F_{alg} behaving as $\hat{\beta}$ for all $i < m$:

$$[\![\beta_i < m \wedge B(i, \beta_i)]\!] = [\![\gamma_i < m \wedge B(i, \gamma_i)]\!]$$

but

$$\langle\!\langle \gamma_i < m \wedge B(i, \gamma_i) \rangle\!\rangle \setminus \langle\!\langle \beta_i < m \wedge B(i, \beta_i) \rangle\!\rangle$$

are non-empty. The only information we have about such difference sets is that their individual counting measures in Ω are infinitesimal.

By a careful choice of $\hat{\beta}$ we need to exclude the possibility that there is an $\alpha \in F$ with $[\![\alpha < m]\!] = 1_{\mathcal{B}}$ defining a distribution on $i < m$ so that the errors of individual β_is combine into a set with standard positive probability.

For any $\hat{\beta}$ of the form as above, define

$$E_i \;:=\; \langle\!\langle \exists y < m B(i,y) \rangle\!\rangle \setminus \langle\!\langle \beta_i < m \wedge B(i, \beta_i) \rangle\!\rangle.$$

Claim 1: *Assume that the counting measure in Ω of $\bigcup_{i<m} E_i \in \mathcal{A}$ is infinitesimal. Then $\Theta \in G_{\text{alg}}$ computed by $\hat{\beta}$ is a Skolem function for $\exists y < m B(x,y)$ on m.*

Assume for the sake of contradiction that for some $\alpha \in F_{\text{alg}}$

$$[\![\Theta(\alpha) < m \wedge B(\alpha, \Theta(\alpha))]\!] < [\![\exists y < m B(\alpha, y)]\!].$$

Applying the observation at the beginning of the section (i.e. Theorem 3.5.2) we get $\gamma \in F_{\text{alg}}$ such that

$$[\![\Theta(\alpha) < m \wedge B(\alpha, \Theta(\alpha))]\!] < [\![\gamma < m \wedge B(\alpha, \gamma)]\!].$$

Therefore

$$U \;:=\; \langle\!\langle \gamma < m \wedge B(\alpha, \gamma) \rangle\!\rangle \setminus \langle\!\langle \Theta(\alpha) < m \wedge B(\alpha, \Theta(\alpha)) \rangle\!\rangle$$

has a non-infinitesimal counting measure in Ω. But clearly

$$U \subseteq \bigcup_{i<m} \left(\{\omega \in \Omega \mid \alpha(\omega) = i\} \cap E_i\right) \subseteq \bigcup_{i<m} E_i,$$

which is a contradiction. This proves the claim.

We will find $\hat{\beta}$ such that the error sets E_i will satisfy the hypothesis of Claim 1. We have

$$\langle\langle \exists y < mB(i,y) \rangle\rangle = \bigcup_{j<m} \langle\langle B(i,j) \rangle\rangle.$$

By Lemma 14.3.1 $\langle\langle B(i,j) \rangle\rangle$ are computed by 0/1-valued trees, elements $\rho_{i,j} \in F_{\mathrm{alg}}$. In fact, there is $t > \mathbf{N}$ such that both the degree and the depth of all $\rho_{i,j}$, $i,j < m$, are bounded above by $n^{1/t}$.

Claim 2: *For all $i,j < m$ there is a polynomial $q_{i,j} \in \mathbf{F}_2^{\mathrm{low}}[x_1,\ldots,x_n]$ of degree bounded by $n^{2/t}$ such that:*

$$\omega \in \langle\langle B(i,j) \rangle\rangle \ \textit{iff} \ q_{i,j}(\omega) = 1.$$

The claim follows from Lemma 14.1.4.

Now comes a key point where we borrow an idea due to Razborov [92]. Fix $r := n^{1/s}$ for some $s > \mathbf{N}$ to be specified later. For each $i < m$ and an arbitrary r-tuple (an element of $\mathcal{M}$) of sets

$$J_{i,1},\ldots,J_{i,r} \subseteq m = \{0,\ldots,m-1\}$$

put:

$$q_i := 1 - \Pi_{v \le r}(1 - \sum_{j \in J_{i,v}} q_{i,j}).$$

The degree of q_i is bounded by $r \cdot n^{2/t} \le n^{(2/t)+(1/s)}$ independently of the choice of sets $J_{i,v}$, so $q_i \in \mathbf{F}_2^{\mathrm{low}}[x_1,\ldots,x_n]$.

Claim 3: *There is a choice (in $\mathcal{M}$) of sets $J_{i,v} \subseteq m$ for $i < m$ and $v \le r$ such that the counting measure of each set,*

$$\langle\langle \exists y < mB(i,y) \rangle\rangle \setminus \{\omega \in \Omega \mid q_i(\omega) = 1\},$$

is at most 2^{-r}.

Work in $\mathcal{M}$. As in Razborov [92] pick the sets independently and uniformly at random from subsets of m. For any particular $\omega \in \langle\langle \exists y < mB(i,y) \rangle\rangle$ the probability that $\sum_{j \in J_{i,v}} q_{i,j}(\omega) \neq 1$ is $1/2$, so the probability that this happens for all

$v \le r$ is 2^{-r}. If one of the sums equals to 1 then $q_i(\omega) = 1$. In other words, the probability of an error on ω is 2^{-r}. An averaging argument yields the required sets $J_{i,v}$.

Polynomial q_i computes (with a small error) whether or not the set $\langle\langle \exists y < mB(i,y) \rangle\rangle$ is non-empty. To find a particular $j < m$ with $\omega \in \langle\langle B(i,j) \rangle\rangle$ we simulate binary search by a depth $\log(m)$ tree β_i, with the query polynomials constructed as in Claim 3. The degree of those polynomials is bounded above by $n^{2/t+1/s}$, so $\beta_i \in F_{\text{alg}}$.

Take $\Theta \in G_{\text{alg}}$ computed by this particular m-tuple $\hat{\beta} = (\beta_0, \ldots, \beta_{m-1})$.

Claim 4: *The function Θ is a Skolem function for $\exists y B(x,y)$ on m.*

By construction the error set

$$\langle\langle \exists y < mB(i,y) \rangle\rangle \setminus \langle\langle \Theta(i) < m \wedge B(i, \Theta(i)) \rangle\rangle$$

has counting measure $\le m \cdot 2^{-r}$ for each $i < m$. Hence the union of these sets has counting measure $\le 2^{-r} m^2$. As $m \in \mathcal{M}_n$, $m^2 \le 2^{n^{1/u}}$ for some $u > \mathbf{N}$. Hence we want to pick r such that:

$$2^{-r} 2^{n^{1/u}} = 2^{n^{1/u} - n^{1/s}}$$

is infinitesimal. Taking $r := n^{1/s}$ for s such that $u > s > \mathbf{N}$ will do. This proves the claim and hence also the lemma. □

15.2 Interpretation of Q_2 in front of an open formula

We shall now interpret the Q_2 quantifier in front of an open formula. That is, for any $m \in \mathcal{M}_n$ and an open $L_n^2(F_{\text{alg}}, G_{\text{alg}})$-formula $A(\overline{x}, y)$ and any $\overline{\alpha}, \beta \in F_{\text{alg}}$ we define the truth value

$$[\![Q_2 y < \beta A(\overline{\alpha}, \beta)]\!].$$

Recall that for an open formula $A(\overline{x}, y)$ and $\overline{\alpha}, \beta \in F_{\text{alg}}$

$$[\![A(\overline{\alpha}, \beta)]\!] = \langle\langle A(\overline{\alpha}, \beta) \rangle\rangle / \mathcal{I}.$$

Moreover, for any $m \in \mathcal{M}_n$ there is a relation $\Gamma \in G_{\text{alg}}$ such that if $\langle\langle \overline{\alpha}, \beta < m \rangle\rangle = \Omega$ then

$$\langle\langle A(\overline{\alpha}, \beta) \equiv \Gamma(\overline{\alpha}, \beta) \rangle\rangle = \Omega.$$

Also, for such $\overline{\alpha}, \beta$,

$$\langle\langle A(\overline{\alpha}, \beta)\rangle\rangle = \bigcup_{\bar{i},j<m} \langle\langle \overline{\alpha} = \bar{i}\rangle\rangle \cap \langle\langle \beta = j\rangle\rangle \cap \langle\langle A(\bar{i}, j)\rangle\rangle.$$

By Lemma 14.1.4 we may assume that Γ is given by an array $(p_{\bar{i},j})_{\bar{i},j<m}$ of polynomials from $\mathbf{F}_2^{\text{low}}[x_1, \ldots, x_n]$:

$$\langle\langle \Gamma(\bar{i}, j)\rangle\rangle = \{\omega \in \Omega \mid p_{\bar{i},j}(\omega) = 1\}.$$

The following definition is the specialization of the notion of 2-cover from Section 13.2 for the model.

Definition 15.2.1 Let $A(\overline{x}, y)$, $m \in \mathcal{M}_n$ and Γ be as above. The **2-cover** of Γ on m is the relation $\Delta \in G$ on m defined by the array:

$$\left(\sum_{k<j} p_{\bar{i},k}\right)_{\bar{i},j<m}.$$

In particular,

$$\langle\langle \Delta(\bar{i}, j)\rangle\rangle = \left\{\omega \in \Omega \mid \sum_{k<j} p_{\bar{i},k}(\omega) = 1\right\}.$$

Definition 15.2.2 Let $A(\overline{x}, y)$ be an open $L_n^2(F_{\text{alg}}, G_{\text{alg}})$-formula and $\overline{\alpha}, \beta \in F_{\text{alg}}$. Take any $m \in \mathcal{M}_n$ such that $\langle\langle \overline{\alpha}, \beta < m\rangle\rangle = \Omega$ and let $\Gamma \in G_{\text{alg}}$ be the relation representing A on m as above. Let Δ be the 2-cover of Γ on m.

Then define:

$$[\![Q_2 y < \beta A(\overline{\alpha}, y)]\!] := [\![\Delta(\overline{\alpha}, \beta)]\!].$$

The following lemma follows directly from the definition; note that, while Γ may not be unique as an array of polynomials, the corresponding polynomials in two such arrays will compute an identical function over $\mathbf{F}_2$.

Lemma 15.2.3 *For any open $L_n^2(F_{\text{alg}}, G_{\text{alg}})$-formula A and any $\overline{\alpha}, \beta \in F_{\text{alg}}$ the definition of $[\![Q_2 y < \beta A(\overline{\alpha}, y)]\!]$ does not depend on the particular choice of $m \in \mathcal{M}_n$ or $\Gamma \in G_{\text{alg}}$.*

Furthermore, the instances of the axioms of the Q_2 quantifier for formula A from Section 13.1 are valid in $K(F_{\text{alg}}, G_{\text{alg}})$.

15.3 Elimination of quantifiers and the interpretation of the Q_2 quantifier

Theorem 15.1.1 provides an elimination of bounded existential or universal quantifiers in front of an open formula, while for the Q_2 quantifier this is achieved simply by its definition (Definition 15.2.2). Doing this in parallel, by induction on the number of quantifiers in a formula, we simultaneously eliminate quantifiers in $Q_2\Sigma^b_\infty\ Q_2L^2_n(F_{\mathrm{alg}}, G_{\mathrm{alg}})$-formulas and define the Q_2 quantifier in front of any $Q_2\Sigma^b_\infty\ Q_2L^2_n(F_{\mathrm{alg}}, G_{\mathrm{alg}})$-formula. Let us summarize this in a lemma suitable for later use.

Lemma 15.3.1 *Let $A(\overline{x})$ be a $Q_2\Sigma^b_\infty\ Q_2L^2_n(F_{\mathrm{alg}}, G_{\mathrm{alg}})$-formula with free variables $\overline{x}$. Let $m \in \mathcal{M}_n$ be arbitrary.*

Then there is an open $L^2_n(F_{\mathrm{alg}}, G_{\mathrm{alg}})$-formula $A'(\overline{x})$ with the same free variables as A such that:

$$[\![\forall \overline{x} < m A(\overline{x}) \equiv A'(\overline{x})]\!] \;=\; 1_{\mathcal{B}}.$$

In other words, $K(F_{\mathrm{alg}}, G_{\mathrm{alg}})$ admits bounded quantifier elimination.

Similarly to Corollary 11.1.3 we note here for future reference a consequence of the construction underlying the quantifier elimination (and the definition of Q_2).

Corollary 15.3.2 *For any $Q_2\Sigma^b_\infty$-formula $A(x_1,\ldots,x_k)$, possibly with parameters from F_{alg} or G_{alg}, and for any $m \in \mathcal{M}_n$ and arbitrarily small but non-standard $s > \mathbf{N}$ there is an open $L^2_n(F_{\mathrm{alg}}, G_{\mathrm{alg}})$-formula $B(x_1,\ldots,x_k)$ such that*

$$[\![\forall x_1,\ldots,x_k < m\ (A(x_1,\ldots,x_k)\ \equiv\ B(x_1,\ldots,x_k))]\!] \;=\; 1_{\mathcal{B}}$$

and such that for any $\alpha_1,\ldots,\alpha_k \in F_{\mathrm{alg}}$ the counting measure of the symmetric difference

$$\left\langle\!\!\left\langle \bigwedge_i \alpha_i < m \rightarrow A(\alpha_1,\ldots,\alpha_k) \right\rangle\!\!\right\rangle \,\triangle\, \left\langle\!\!\left\langle \bigwedge_i \alpha_i < m \rightarrow B(\alpha_1,\ldots,\alpha_k) \right\rangle\!\!\right\rangle$$

is less than $2^{-n^{1/s}}$. In particular, it is infinitesimal.

15.4 Comprehension and induction for $Q_2\Sigma_0^{1,b}$-formulas

The model $K(F_{\rm alg}, G_{\rm alg})$ admits bounded quantifier elimination (Lemma 15.3.1), and open comprehension and open induction are valid in it (Lemma 14.3.2). Hence Theorem 9.3.1 yields the following.

Theorem 15.4.1 *Comprehension and induction for $Q_2\Sigma_0^{1,b}$ $L_n(F_{\rm alg}, G_{\rm alg})$-formulas are valid in $K(F_{\rm alg}, G_{\rm alg})$.*

In particular, theory $Q_2V_1^0$ is valid in $K(F_{\rm alg}, G_{\rm alg})$.

16
Witnessing and independence in $Q_2V_1^0$

We have considered witnessing theorems, independence results and definability in theory V_1^0 in Chapter 12. As an example we have proved there a witnessing theorem for $\forall X < x\exists Y < x\Sigma_0^{1,b}$-formulas. The proof proceeds by analyzing model $K(F_{\text{tree}}, G_{\text{tree}})$. It should be clear that structure $K(F_{\text{alg}}, G_{\text{alg}})$ can be used for an essentially identical proof to provide a quite analogous witnessing of $\forall X < x\exists Y < x\Sigma_0^{1,b}$-formulas in $Q_2V_1^0$.

We shall prove in Section 16.1 a witnessing theorem for the second case considered in the introduction to Chapter 12, the $\forall X < x\exists Y < x\forall Z < x\Sigma_0^{1,b}$-formulas. Again it should be obvious that a similar argument applies to V_1^0 and $K(F_{\text{tree}}, G_{\text{tree}})$ as well.

In Section 16.2 we note that the proofs of the preservation theorems for $K(F_{\text{tree}}, G_{\text{tree}})$ from Section 12.2 apply also to $K(F_{\text{alg}}, G_{\text{alg}})$.

16.1 Witnessing $\forall X < x\exists Y < x\forall Z < x\Sigma_0^{1,b}$-formulas

Let $A(x)$ be a $\forall X < x\exists Y < x\forall Z < x\Sigma_0^{1,b}$-formula of the form

$$\forall X < x\exists Y < x\forall Z < xB(x, X, Y, Z)$$

with B a $\Sigma_0^{1,b}$-formula. Families F_{alg} and G_{alg} are closed under definitions by open formulas (Lemma 14.3.1) and so we may apply Lemma 3.3.3 to deduce the following statement for $K(F_{\text{alg}}, G_{\text{alg}})$.

Lemma 16.1.1 *Assume* $[[A(n)]] = 1_{\mathcal{B}}$. *Then for every* $\Delta \in G_{\text{alg}}$ *and every standard* $\epsilon > 0$ *there is* $\Gamma \in G_{\text{alg}}$ *such that for every* $\Theta \in G_{\text{alg}}$*:*

$$\mu([[\Delta < n \to (\Gamma < n \wedge (\Theta < n \to B(n, \Delta, \Gamma, \Theta)))]]) > 1 - \epsilon.$$

Let us take for Δ the identity on $\Omega = \{0,1\}^n$, $\Delta(\omega) = \omega$. Applying Corollary 15.3.2 to $B(n,\Delta,\Gamma,\Theta)$ and then Lemma 2.2.1 we obtain the next statement.

Theorem 16.1.2 *Assume $[\![A(n)]\!] = 1_{\mathcal{B}}$. Then for every standard $\epsilon > 0$ there is $\Gamma \in G_{\text{alg}}$ such that for all $\Theta \in G_{\text{alg}}$*

$$\text{Prob}_{(\omega\in\Omega}[\ \omega \in \langle\langle \Gamma < n \wedge (\Theta < n \rightarrow B(n,\Delta,\Gamma,\Theta))\rangle\rangle\ ] > 1-\epsilon.$$

In particular, this holds if theory $Q_2V_1^0$ proves $\forall x A(x)$.

We may derive a finitary version of this theorem analogously to Theorem 12.1.4, using compactness and the fact that we have not used any property of n except that it is non-standard.

Corollary 16.1.3 *Assume that theory $Q_2V_1^0$ proves $\forall x A(x)$. Then for every standard $\epsilon,\delta > 0$ and $m \in \mathbf{N}$ there is a collection $p = (p_i)_{i<m}$ of polynomials over $\mathbf{F}_2[x_0,\ldots,x_{m-1}]$ of degree less than n^δ such that for any other collection $q = (q_i)_{i<m}$ of polynomials over $\mathbf{F}_2[x_0,\ldots,x_{m-1}]$ of degree less than n^δ the following holds for all m large enough:*

$$\text{Prob}_{\omega\in\{0,1\}^m}[\ B(m,\omega,p(\omega),q(\omega))\ ] > 1-\epsilon.$$

Here $p(\omega)$ abbreviates $(p_i(\omega))_{i<m}$.

We now give an example of an independence result for a formula of the quantifier complexity considered in the witnessing theorem. We will formulate it as a statement about $K(F_{\text{alg}}, G_{\text{alg}})$ (and use Theorem 16.1.2) rather than a finitary statement. But it will be clear that it is possible to use in place of n a number parameter, getting a finitary statement (with a finitary independence proof in which the role of Theorem 16.1.2 is taken up by Corollary 16.1.3).

We shall assume n has the form $n = \binom{m}{2}$, for some $m \in \mathcal{M}_n$, and we shall interpret subsets of $[n]$ as defining undirected graphs on $[m]$ (without loops). In particular, the sample space Ω is now interpreted as a set of graphs X on m vertices. With this understanding we shall write $(u,v) \in X$ for $u,v \in [m]$, meaning that the edge (u,v) is in graph X.

Let $A(n,X)$ be a $\Sigma_0^{1,b}$-formula

$$X \subseteq [n]\ \wedge\ \forall u,v,w \in [m]\ (((u,v) \in X \wedge (v,w) \in X) \rightarrow (u,w) \in X).$$

The formula says that X is transitive.

The true $\forall X \exists Y \forall Z \Sigma_0^{1,b}$ sentence which fails in $K(F_{\mathrm{alg}}, G_{\mathrm{alg}})$ (and hence is independent of $Q_2V_1^0$) asserts that any graph has a transitive closure:

$$\forall X \exists Y \forall Z\, [\, X \subseteq Y \wedge A(n,Y) \wedge ((X \subseteq Z \wedge A(n,Z)) \rightarrow Y \subseteq Z)\,]$$

(we omitted the bound $\subseteq [n]$ for X, Y, Z). Denote this sentence Closure(n).

Theorem 16.1.4 *In $K(F_{\mathrm{alg}}, G_{\mathrm{alg}})$ it holds that* $[\![\mathrm{Closure}(n)]\!] = 0_{\mathcal{B}}$.

In particular, $Q_2V_1^0$ does not prove the existence of a transitive closure of a graph.

We will not give a proof of the theorem here using the witnessing theorem, as a different but simpler one can be given. The statement is included as an example of a sentence of the right quantifier complexity.

16.2 Preservation of true $s\Pi_1^{1,b}$-sentences

In Section 12.2 we showed that all true sentences $\forall x A(x)$ are valid in the model $K(F_{\mathrm{tree}}, G_{\mathrm{tree}})$, as long as formula $A(x)$ has one of the following three forms:

(c) $\forall X < x \Sigma_0^{1,b}$ (i.e. $s\Pi_1^{1,b}$),
(d) $\forall X < x \exists y < x \forall Z < x \Sigma_0^{1,b}$,
(e) $\exists Y < x \forall Z < x \Sigma_0^{1,b}$.

The key case was (c); the preservation in the other two cases are its corollaries. The preservation for case (c) (Theorem 12.2.1) was proved using Corollary 11.1.3, a consequence of the proof of bounded quantifier elimination for $K(F_{\mathrm{tree}}, G_{\mathrm{tree}})$. For structure $K(F_{\mathrm{alg}}, G_{\mathrm{alg}})$ we have the analogous statement in Corollary 15.3.2, and the same argument works.

The preservation for cases (d) and (e) were then derived using a collection scheme in $\mathcal{M}$ and the richness of language L_n respectively (see Corollaries 12.2.2 and 12.2.3), and identical arguments apply for $K(F_{\mathrm{alg}}, G_{\mathrm{alg}})$. We shall thus only state the theorem.

Theorem 16.2.1 *Assume sentence $\forall x A(x)$ is true where $A(x)$ has one of the three forms (c), (d) or (e) listed above. Then $\forall x A(x)$ is valid in $K(F_{\mathrm{alg}}, G_{\mathrm{alg}})$.*

PART V

Towards proof complexity

17
Propositional proof systems

In this part we survey a few basic definitions and facts from proof complexity and model theory of bounded arithmetic. The material discussed in this and the next two chapters is mostly well known and the reader may find a lot more in the literature suggested in the Introduction. In particular, the known material mentioned in this chapter, and indeed in the whole of Part V, can be found in Krajíček[56].

We also present only the definitions and the facts we shall use later, not other related material, and we do not include specific examples illustrating general definitions. Although the exposition is brief, it is fairly complete and should be sufficient for understanding the arguments in the later parts of the book.

We start in this chapter by introducing the proof systems we shall be concerned with in later chapters. The **DeMorgan language** for propositional logic has constants 1 (true) and 0 (false), unary connective $\neg$ (negation) and binary connectives $\vee$ (disjunction) and $\wedge$ (conjunction). We shall tacitly assume that the language of any proof system defined in this chapter contains the DeMorgan language.

17.1 Frege and Extended Frege systems

Definition 17.1.1 (Cook–Reckhow [31]) A **Frege rule** is a $(k+1)$-tuple of formulas $A_0,\dots,A_k$ in atoms $p_1,\dots,p_n$ written as:

$$\frac{A_1,\dots,A_k}{A_0},$$

such that any truth assignment $a:\{p_1,\dots,p_n\}\to\{0,1\}$ satisfying all formulas $A_1\dots,A_k$ satisfies also A_0 (this condition is called the soundness of the rule). A Frege rule in which $k=0$ is called a **Frege axiom scheme**.

An instance of the rule is obtained by a simultaneous substitution of arbitrary formulas B_i for all p_i.

Definition 17.1.2 (Cook–Reckhow [31]) Let F be a finite collection of Frege rules. A **Frege proof** (an F-proof for short) of formula D from formulas $C_1,\ldots,C_u$ is a finite sequence $B_1,\ldots,B_k$ of formulas such that $B_k = D$, and such that every B_i is either one of $C_1,\ldots,C_u$, or is inferred from some earlier B_js $(j < i)$ by a rule of F.

F is implicationally complete if and only if any D can be F-proved from any set $\{C_1,\ldots,C_u\}$ if every truth assignment satisfying all C_is satisfies also D.

F is a **Frege proof system** if and only if it is implicationally complete.

The next class of proof systems augments Frege systems by the ability to abbreviate formulas by new atoms. We use the connective $\equiv$ in its definition. If $\equiv$ is not in the language of F we use any fixed DeMorgan formula defining it.

Definition 17.1.3 (Cook–Reckhow [31]) Let F be a Frege system. An **extended Frege proof** is a sequence of formulas $A_1,\ldots,A_k$ such that every A_i is either obtained from some previous A_js by an F-rule or has the form:

$$q \equiv B$$

with the following conditions satisfied:

1. Atom q does not appear in B, or in any A_j for $j < i$.
2. Atom q does not appear in A_k.

A formula of this form is called an **extension axiom**; q is called an **extension atom**.

An **Extended Frege system**, EF, is a proof system whose proofs are extended Frege proofs.

The ability to introduce an extension axiom in EF is sometimes called the 'Extension rule' but it is not a Frege rule in the earlier sense.

It is a well-known fact that the definition of Frege systems is very robust. As far as lengths of proofs are concerned the strength of a Frege system does not depend on its language or particular format of proofs (from many variants considered in proof theory).This means that a proof of a DeMorgan formula in one Frege system can be translated (by a polynomial-time algorithm even) into an at most polynomially longer proof of the same formula in any other Frege system. The same relations hold among different Extended Frege systems.

17.2 Language with connective $\oplus$ and constant-depth Frege systems

The next definition is a particular case of a more general definition extending DeMorgan language by modular counting connectives $MOD_{a,i}$; see Krajíček [56] or Buss *et al.* [18].

Definition 17.2.1 $\oplus$ is a propositional connective of unbounded arity such that

$$\oplus(p_1,\ldots,p_k) \text{ is true} \quad \text{iff} \quad |\{j \mid p_j \text{ true}\}| \textit{ is odd.}$$

The $\oplus$-axioms are the formula

$$\neg \oplus (\emptyset)$$

together with the axiom schemes (one for each $k \geq 1$)

$$\oplus(A_1,\ldots,A_k) \equiv$$

$$[(\oplus(A_1,\ldots,A_{k-1}) \wedge \neg A_k) \vee (\neg \oplus (A_1,\ldots,A_{k-1}) \wedge A_k)].$$

We shall study Frege systems and their subsystems that have either the DeMorgan language or the DeMorgan language with $\oplus$. Thus we shall make the provision that for a Frege system in the DeMorgan language we use the name F while a Frege system in the DeMorgan language augmented by the $\oplus$ connective will be named $F(\oplus)$.

This extension of the language is not so interesting for unrestricted Frege systems as there is a quadratic size DeMorgan formula defining $\oplus$, and F polynomially proves the translation of the $\oplus$-axioms.

However, having $\oplus$ in the language makes a substantial difference if proofs are restricted to formulas of bounded depth. This is because any constant-depth DeMorgan formula defining parity of n bits must have the size exponential in n; see Theorem 12.3.1 of Ajtai [1], Furst *et al.* [34], Yao [108] and Hastad [40].

The restriction to constant depth means that only a constant number of alternations between disjunctions and conjunctions are allowed in any formula; the depth of propositional formulas is analogous to the quantifier complexity of first-order formulas. One may define Frege systems as operating with conjunctions and disjunctions of unbounded arity; the depth is then simply the depth of the formula. However, it seems easier to define the depth of a formula in a less straightforward way rather than modifying the definition of Frege systems.

The unbounded $\bigvee$ and $\bigwedge$ we sometimes use in formulas should be understood as being built from binary $\vee$ and $\wedge$, respectively.

Definition 17.2.2 The depth of a formula A in the DeMorgan language augmented by $\oplus$, denoted $\mathrm{dp}(A)$, is defined inductively as follows:

1. The depth of constants and atoms is 0.
2. $\mathrm{dp}(\neg A) := 1 + \mathrm{dp}(A)$.
3. $\mathrm{dp}(\oplus(\emptyset)) := 0$ and
   $\mathrm{dp}(\oplus(A_1, \ldots, A_k)) := 1 + \max_{i \leq k} \mathrm{dp}(A_i)$.
4. If $B(q_1, \ldots, q_k)$ is a formula built from atoms using only $\vee$ and if none of the formulas $A_1, \ldots, A_k$ has $\vee$ as the top connective, then for $C = B(A_1, \ldots, A_k)$ we put:

$$\mathrm{dp}(C) := 1 + \max_{i \leq k} \mathrm{dp}(A_i),$$

   and analogously for $\wedge$.

One can get a somewhat cleaner definition by requiring that $\neg$ is applied only to atoms, and giving all literals depth 0. However, we will be concerned with proofs that use only bounded depth formulas, but not in the particular bounds on the depth, and hence such nuances of the definition are irrelevant for our purposes.

Note that DNF formulas have depth at most 3. The lower bound $d \geq 3$ in the next definition is to guarantee that the set of depth d DeMorgan tautologies is coNP-complete.

Definition 17.2.3 For fixed $d \geq 3$, F_d is the subsystem of a Frege system F in the DeMorgan language whose proofs may contain only formulas of depth at most d.

Similarly, $F_d(\oplus)$ is the subsystem of a Frege system $F(\oplus)$ in the DeMorgan language augmented by $\oplus$ whose proofs may contain only formulas of depth at most d.

Note that it seems that we cannot directly compare the strength of, say, F and F_d as the latter proof system cannot in principle prove tautologies of depth higher than d. Similarly, comparing F with $F(\oplus)$ appears cumbersome as the class of tautologies in the language with $\oplus$ is bigger than in the DeMorgan language only.

In fact, these and many other issues that come up when developing a theory of proof systems can be successfully addressed. In particular, there is a general definition of a *propositional proof system* that subsumes the definitions of Frege systems and Extended Frege systems and their depth d fragments.

We will not explain this material here; the reader may find it in details in Cook and Reckhow [31] or in Krajíček[56], or briefly in Part VIII.

18

An approach to lengths-of-proofs lower bounds

We now outline a general strategy for proving lower bounds for lengths-of-proofs, showing that it is sufficient to construct a certain model of bounded arithmetic. As we aim for a stronger proof system, we need to construct a model of a stronger theory. This reduction uses essentially just the compactness of first-order logic and provability of the soundness of a proof system in the corresponding bounded arithmetic theory, and goes back to Ajtai [2]. Here we describe this approach, modified to allow our Boolean-valued structures.

18.1 Formalization of the provability predicate

Let us start with a bit more general discussion. Assume that for every $k \in \mathbf{N}$ there is a string w_k of length at most $s(k)$, where function $s(k)$ is subexponential:

$$s(k) < 2^{k^{o(1)}}.$$

The string can be encoded by a set $W_k \subseteq s(k)$ and the whole sequence $\{W_k\}_{k \in \mathbf{N}}$ by a binary relation $W(x,y)$ such that

$$W_k = \{m \in \mathbf{N} \mid W(k,m)\}.$$

Assume that $P(X)$ is an NP-property of strings (finite sets). By Fagin's theorem $P(X)$ can be defined by an $s\Sigma_1^{1,b}$ L_n-formula

$$\exists Y A(X,Y)$$

where A is a $\Sigma_0^{1,b}$-formula, with the quantifiers of A and the length of the string Y bounded polynomially in terms of the length of the string X (we can take just unary Y as there is a pairing function in L_n). Hence there is a binary relation $V(x,y)$ such that each

$$V_k = \{m \in \mathbf{N} \mid V(k,m)\}$$

is a subset of $s(k)^{O(1)}$ and whenever W_k has the property $P(X)$ then V_k witnesses the validity of $P(W_k)$: $A(W_k, V_k)$ holds (language L_n has names for W and V, so this is a sentence of the language).

Assume now that W_k has the property $P(X)$ for infinitely many $k \in \mathbf{N}$. By elementariness of $\mathcal{M}$ it follows that W_n also satisfies $P(X)$ for some non-standard $n \in \mathcal{M}$ (in fact, for cofinaly many of them in $\mathcal{M}$). Similarly, if W_k had the property $P(X)$ for all $k \in \mathbf{N}$, W_n would also satisfy $P(X)$ for all $n \in \mathcal{M}$.

Now we apply this general argument to the provability predicate. Let P denote any of the proof systems defined earlier: F, $F(\oplus)$, EF, F_d or $F_d(\oplus)$.

Assume that $T(x,y)$ and $R(x,y)$ are two binary relations such that for all $k \in \mathbf{N}$ both sets

$$T_k = \{m \in \mathbf{N} \mid T(k,m)\}$$

and

$$R_k = \{m \in \mathbf{N} \mid R(k,m)\}$$

are subsets of $s(k)$ for some subexponential function $s(k) < 2^{k^{0(1)}}$, and such that for some $d \geq 1$ and for infinitely many $k \in \mathbf{N}$ the following hold:

1. T_k encodes a DeMorgan formula of depth at most d, and
2. R_k encodes a P-proof of T_k.

Then for some non-standard $n \in \mathcal{M}$ the statement

$$(*) \qquad R_n \text{ is a } P\text{-proof of a depth } d \text{ formula } T_n$$

is true in $\mathcal{M}$. This follows from the general discussion as the properties of being a DeMorgan formula or a P-proof are polynomial-time properties and can be defined in the same way as above.

In fact, this general argument referring to Fagin's theorem is not quite adequate for our purpose as one needs to be able to prove various properties of these predicates. What we need is that the predicates can be defined following the usual syntax of logic and in a way that their basic properties hold and are provable in a weak bounded arithmetic theory. It is well known how such a formalization of logic can be developed (see e.g. Krajíček [56, 9.3]) and we shall not repeat it here. Further we may assume without loss of generality that the encoding of proofs comes with an auxiliary structure (a witness to the $s\Sigma_1^{1,b}$-formula) so that (*) is expressible as

$$\mathrm{Fla}_d(n,T_n) \wedge \mathrm{Prf}_P(n,R_n,T_n),$$

a conjunction of instances of two $\Sigma_0^{1,b}$-formulas

$$\mathrm{Fla}_d(x,Y)$$

defining depth d formulas and

$$\mathrm{Prf}_P(x,X,Y)$$

defining the provability predicate. In these formulas X and Y are bounded by an L_n-term in x; function $s(k)$ was subexponential in k and hence there is a function symbol in L_n for it. In particular, both R_n and T_n are subsets of $\mathcal{M}_n$.

Lemma 18.1.1 *For any non-standard $n \in \mathcal{M}$ satisfying (*) the statement*

$$\mathrm{Fla}_d(n,T_n) \wedge \mathrm{Prf}_P(n,R_n,T_n)$$

is valid in any L_n-closed structure $K(F,G)$.

Proof: The statement is an instance of a, $\forall\Sigma_\infty^{1,b}$-sentence true in $\mathcal{M}$. Hence it is valid in any L_n-closed structure, analogously with Lemma 5.5.1. □

18.2 Reflection principles

To formalize a reflection principle for a proof system P we need a formula defining the satisfiability relation. For general formulas this can be defined by both $s\Pi_1^{1,b}$ and $s\Sigma_1^{1,b}$-formulas but not by a $\Sigma_0^{1,b}$-formula. When the depth of formulas is bounded above by some standard d we can construct a $\Sigma_0^{1,b}$ definition

$$\mathrm{Sat}_d(x,Z,Y).$$

This says that a truth assignment Z satisfies a depth $\leq d$ formula Y, with x bounding both Z and Y.

The construction of such a formula can be done in a way that V_1^0 proves its basic properties corresponding to Tarski's conditions for a truth definition. This can be found in Krajíček[56].

Definition 18.2.1 For $d \geq 1$ define the formula

$$\mathrm{Ref}_{P,d}(x,X,Y,Z)$$

to be the implication

$$(\mathrm{Fla}_d(x,Y) \wedge \mathrm{Prf}_P(x,X,Y)) \to \mathrm{Sat}_d(x,Z,Y),$$

Ref_P denotes the universal closure of $\mathrm{Ref}_{P,3}(x,X,Y,Z)$.

Parameter d is chosen to be 3 in the definition of Ref_P as DNF tautologies have depth at most 3 all proof systems considered so far can operate with depth 3 formulas and also all hard tautologies we deal with are DNF.

Sentence Ref_P is the universal closure of a $s\Pi_1^{1,b}$-formula and it is true (for the proof systems P we have defined). Hence we know, by Theorems 12.2.1 and 16.2.1, that it is valid in structures $K(F_{\mathrm{tree}}, G_{\mathrm{tree}})$ and $K(F_{\mathrm{alg}}, G_{\mathrm{alg}})$. However, for proof complexity it is key that this validity also follows differently.

Theorem 18.2.2 *Theory V_1^0 proves $\mathrm{Ref}_{F_d,d}$ and theory $Q_2V_1^0$ proves $\mathrm{Ref}_{F_d(\oplus),d}$, for all $d \geq 3$.*

This theorem is well known and we shall not reprove it here. The idea behind its proof is that, given a truth assignment Z, one verifies by induction on the number of steps ℓ in a depth d proof X of a formula Y that the ith step in X is satisfied by Z, for $i = 1, \ldots, \ell$.

18.3 Three conditions for a lower bound

Now we put the ideas from the preceding sections together.

Theorem 18.3.1 *Let $d \geq 3$ and let T_k be depth d tautologies of size $k^{O(1)}$, for all $k \in \mathbf{N}$. Further let P be any of the proof systems of Chapter 17.*

Assume that for an arbitrary choice of non-standard n there is a structure $K(F,G)$ (built over $\mathcal{M}_n$) satisfying the following three conditions:

(1) $K(F,G)$ is L_n-closed.
(2) $\mathrm{Ref}_{P,d}$ is valid in $K(F,G)$.
(3) There is a truth assignment $\Gamma \in G$ to atoms of T_n that falsifies the formula in $K(F,G)$, i.e. $\mathrm{Sat}_d(n,\Gamma,\neg T_n)$ is valid.

Then there is a standard $\epsilon > 0$ such that for no $k \in \mathbf{N}$ large enough has T_k a P-proof of size less than 2^{k^ϵ}.

Proof: We give a brief sketch of the proof (which is well known).

Assume for the sake of a contradiction that for infinitely many $k \in \mathbf{N}$ there are P-proofs R_k of T_k of a subexponential size $2^{k^{o(1)}}$. As in Section 18.1, there is a non-standard n such that (for relations R and T with names in L_n) statement

$$\mathrm{Fla}_d(n, T_n) \wedge \mathrm{Prf}_P(n, R_n, T_n)$$

is true in $\mathcal{M}$. Hence by Lemma 18.1.1 and condition (1) it is also valid in $K(F,G)$.

It follows by condition (2) that

$$\forall Z \mathrm{Sat}_d(n, Z, T_n)$$

is valid in $K(F,G)$. But by condition (3) $\mathrm{Sat}_d(n, \Gamma, \neg T_n)$ is valid for some $\Gamma \in G$. It follows by properties of Sat_d corresponding to Tarski's conditions that also

$$\neg \mathrm{Sat}_d(n, \Gamma, T_n)$$

is valid. But that is a contradiction. □

Hence, in order to prove that no R_n could exist in $\mathcal{M}_n$ and thus to prove an exponential lower bound to the size of P-proofs of formulas T_k, $k \in \mathbf{N}$, it suffices to construct a structure $K(F,G)$ satisfying the three conditions. In fact, it would suffice to have $[[\mathrm{Sat}_d(n, \Gamma, T_n)]] < 1_{\mathcal{B}}$.

19

PHP principle

Many combinatorial principles (and, more generally, statements about finite structures) can be formalized both in first-order logic and in propositional logic. They are very often expressible by $\Sigma_0^{1,b}$-formulas, and instances of these can be translated into propositional ones quite easily (replacing bounded existential and universal quantifiers by big disjunctions or conjunctions). In fact, the propositional translation can be defined for more general formulas, including any of the three forms (c), (d) and (e) defined in the introduction to Chapter 12. In this translation a universally true formula is translated into a sequence of tautologies, one tautology for every instance of the universally quantified number variable.

The topic of propositional translations is very well developed and quite subtle in places. We shall not give a general definition but simply present the formalization for the **pigeonhole principle** (PHP) only. The reader may find detailed expositions in Krajíček[56] and in Cook and Nguyen [30].

19.1 First-order and propositional formulations of PHP

Let $R(x,y)$ be a second-order variable for a binary relation symbol. The $\Sigma_0^{1,b}$-formula $\mathrm{PHP}(x,R)$ is the disjunction of the following three formulas:

1. $\exists u \in [x+1] \forall v \in [x] \neg R(u,v)$.
2. $\exists v,v' \in [x] \exists u \in [x+1]\ (v \neq v' \wedge R(u,v) \wedge R(u,v'))$.
3. $\exists u,u' \in [x+1] \exists v \in [x]\ (u \neq u' \wedge R(u,v) \wedge R(u',v))$.

A relation R violating the first two properties for some $x := k$ is a graph of a map of $[k+1]$ into $[k]$, and if R also violated the third condition it would be an injective map. But that is impossible. Hence $\forall x \forall R\ \mathrm{PHP}(x,R)$ is a true statement.

Now we formulate PHP propositionally.

Definition 19.1.1 Let $k \geq 1$ be arbitrary. For any $i \in [k+1]$ and $j \in [k]$, let p_{ij} be a propositional variable. The formula PHP_k, formed from atoms p_{ij}, is the following one:

$$\bigvee_{i \in k+1} \bigwedge_{j \in k} \neg p_{ij} \vee \bigvee_{i \in k+1} \bigvee_{j_1 \neq j_2 \in k} (p_{ij_1} \wedge p_{ij_2}) \vee \bigvee_{i_1 \neq i_2 \in k+1} \bigvee_{j \in k} (p_{i_1 j} \wedge p_{i_2 j}).$$

Again a falsifying assignment $p_{ij} := a_{ij} \in \{0,1\}$ to PHP_k would define the graph of an injective function from $[k+1]$ into $[k]$. Hence formulas PHP_k, $k \geq 1$, are tautologies. Also note that these are DNF formulas of depth 3.

A landmark result in proof complexity is Ajtai's proof [2] that formulas PHP_k cannot be proved by polynomial-size proofs in constant-depth Frege systems F_d (in DeMorgan language). He proved it using the approach of Chapter 18, constructing a suitable model (classical, not Boolean valued) of $I\Delta_0(R)$ where PHP fails for some n. The lower bound has been subsequently strengthened to an exponential one, see Krajíček *et al.* [79] and Pitassi *et al.* [88].

Theorem 19.1.2 (Ajtai [2], Krajíček *et al.* [79], Pitassi *et al.* [88]) *Theories $I\Delta_0(R)$ and V_1^0, even augmented by any function of a subexponential growth, do not prove $\forall x \mathrm{PHP}(x,R)$.*

No constant-depth Frege system F_d (in the DeMorgan language) proves tautologies PHP_k, $k \geq 1$, in a subexponential size.

In the next part we shall give another construction of a suitable model implying the lower bound. Although a form of a switching lemma is used in both constructions there is a subtle difference. In Ajtai[2] one expands by (essentially) a model-theoretic forcing a cut in $\mathcal{M}$ by a new relation $R \subseteq [n+1] \times [n]$ and no new numbers are added (i.e. no new pigeons or new holes). In our construction the relation is a specific random variable known in advance but we add many new pigeons and new holes that were not in $\mathcal{M}_n$.

19.2 Three conditions for F_d and $F_d(\oplus)$ lower bounds for PHP

We now formulate the model-theoretic criterion for lower bounds for the constant-depth proof systems and for the particular formulas PHP_k only.

Theorem 19.2.1 *Let $d \geq 3$ be arbitrary. Assume that for an arbitrary choice of a non-standard n there is a structure $K(F,G)$ (built over $\mathcal{M}_n$) satisfying the following three conditions:*

(1) $K(F,G)$ is L_n-closed.
(2) Theory V_1^0 is valid in $K(F,G)$.
(3) There is a truth assignment $\Gamma \in G$ to atoms of PHP_n *that falsifies* PHP_n *in $K(F,G)$, i.e.* $\text{Sat}_3(n,\Gamma,\neg\text{PHP}_n)$ *is valid.*

Then there is a standard $\epsilon > 0$ such that for no $k \in \mathbf{N}$ large enough has PHP_k *an F_d-proof of size less than 2^{k^ϵ}.*

If a structure exists satisfying conditions (1), (3) and the condition

(2′) Theory $Q_2V_1^0$ is valid in $K(F,G)$.

then there is a standard $\epsilon > 0$ such that for no $k \in \mathbf{N}$ large enough has PHP_k *an $F_d(\oplus)$-proof of size less than 2^{k^ϵ}.*

Proof: The theorem follows from Theorem 18.3.1, using Theorem 18.2.2 to replace condition (2) there by the conditions that the structure is a model of V_1^0 or $Q_2V_1^0$, respectively. □

We should now realize there are troubles lying ahead. Earlier we built the tree model $K(F_{\text{tree}}, G_{\text{tree}})$ and the algebraic model $K(F_{\text{alg}}, G_{\text{alg}})$ with the hope of mimicking their constructions and defining a structure $K(F,G)$ satisfying the conditions in the theorem. But the structure $K(F,G)$, unlike the tree model and the algebraic model, cannot preserve true $s\Pi_1^{1,b}$-sentences as PHP is one of them and we want to violate it. This is actually one of the reasons why we have analyzed the validity of induction in $K(F_{\text{tree}}, G_{\text{tree}})$ and $K(F_{\text{alg}}, G_{\text{alg}})$ in the explicit way we did; a simpler verification would note that the induction axioms are $s\Pi_1^{1,b}$-formulas and appeal to the preservation property.

A structure $K(F,G)$ useful for lengths-of-proofs lower bounds cannot be an $s\Pi_1^{1,b}$-elementary extension of $\mathcal{M}_n$.

PART VI

Proof complexity of F_d and $F_d(\oplus)$

20
A shallow PHP model

We will define a second-order structure $K(F_{\mathrm{PHP}}, G_{\mathrm{PHP}})$ in Chapter 21 in which V_1^0 will be valid but the sentence $\mathrm{PHP}(n,R)$ will fail (will have the truth-value $0_{\mathcal{B}}$) for a particular interpretation of R by an element of G_{PHP}. Theorem 19.1.2 will follow as a corollary of Theorem 19.2.1.

Structure $K(F_{\mathrm{PHP}}, G_{\mathrm{PHP}})$ is constructed in an analogous way to the structure $K(F_{\mathrm{tree}}, G_{\mathrm{tree}})$ but it is based on a different concept of a tree: the PHP-tree. In this chapter we introduce this key concept and investigate its properties using a structure $K(F^0_{\mathrm{PHP}}, G^0_{\mathrm{PHP}})$, a rudimentary version of future $K(F_{\mathrm{PHP}}, G_{\mathrm{PHP}})$, based on shallow PHP-trees.

20.1 Sample space Ω^0_{PHP} and PHP-trees

We begin with the definition of the sample space for the shallow tree model.

Definition 20.1.1 The sample space Ω^0_{PHP} consists of all $\omega \subseteq [n+1] \times [n]$ such that

1. ω is a graph of a partial injective function from $[n+1]$ into $[n]$, and
2. $|\omega| = n$ (i.e. ω is surjective).

We shall often write $\omega(i,j)$ or $\omega(i) = j$ instead of $(i,j) \in \omega$.

In the next section we plan to interpret R by the identity function on Ω^0_{PHP}; this element of the future G^0_{PHP} will be denoted Δ^0. That is, Δ^0 is the random variable

$$\Delta^0 : \omega \in \Omega^0_{\mathrm{PHP}} \to \Delta^0_\omega := \omega.$$

In other words, for elements α, β of future F^0_{PHP} the atomic statement $R(\alpha, \beta)$ is interpreted by $\Delta^0(\alpha) = \beta$, and it holds at sample ω iff

$$\omega(\alpha(\omega)) \ = \ \beta(\omega).$$

We wish to define $K(F^0_{\text{PHP}}, G^0_{\text{PHP}})$ such that $\text{PHP}(n, \Delta^0)$ will fail in the structure, in fact:

$$[\![\text{PHP}(n, \Delta^0)]\!] \ = \ 0_{\mathcal{B}}.$$

Note that the chosen interpretation of R means that we cannot simply take for F^0_{PHP} the cut $\mathcal{M}_n$. This is because for any $i \in [n+1]$ and $j \in [n]$ the equality $\Delta^0(i) = j$ holds for a fraction $\frac{1}{n+1}$ of samples ω (i has n options of where to be mapped by ω and also the option of being outside the domain of ω) and hence

$$[\![\Delta^0(i) = j)]\!] = 0_{\mathcal{B}}.$$

In other words, we need to add new holes where pigeons $i \in [n+1]$ will map to, and this means also new pigeons.

These considerations lead to the following auxiliary concept of a PHP-tree, taken from Beame *et al.* [7]. Families F^0_{PHP} and G^0_{PHP} will be defined using this notion in a way quite analogous to $K(F_{\text{rud}}, G_{\text{rud}})$.

Definition 20.1.2 A **PHP-tree** T is a finite tree whose inner nodes are labeled either by

$$i \to ?$$

for some $i \in [n+1]$, or by

$$? \to j$$

for some $j \in [n]$. If a node is labeled by $i \to ?$ it has n outgoing edges labeled by all $j \in [n]$: edge j corresponds to the case when $\omega(i,j)$ holds. If a node is labeled by $? \to j$ it has $(n+1)$ outgoing edges labeled by all $i \in [n+1]$, with edge i corresponding to the case $\omega(i,j)$.

Any $\omega \in \Omega^0_{\text{PHP}}$ determines a unique path in T. This path, however, need not finish in a leaf. It may happen that a query $i \to ?$ is posed along the path such that $i \notin \text{dom}(\omega)$. If a leaf is reached we shall denote it $T(\omega)$, otherwise $T(\omega)$ is undefined.

The depth $\text{dp}(T)$ of T is defined as always: the maximal number of edges on a path from the root to a leaf.

Lemma 20.1.3 *Let T be a PHP-tree of depth d. Then*

$$\text{Prob}_{\omega \in \Omega^0_{\text{PHP}}}[T(\omega) \ \textit{is undefined}] \ \leq \ O(d/n).$$

Proof: The lemma is proved by induction on n and the depth d of T. The phrase 'induction on n' means the following: we have defined PHP-trees using n as the a priori parameter. But we could equally well have used another parameter m, starting with the sample space consisting of partial bijections between $[m+1]$ and $[m]$ of size m, yielding 'PHP-trees over m'. The induction argument implies the lemma using a similar statement with $n-1$ and $d-1$ in place of n and d.

In particular, denote by $p(d,m)$ the probability (as computed in $\mathcal{M}$) that T is undefined on a sample (also over m) for PHP-trees of depth d over m. Then clearly:

$$p(1,m) = \frac{1}{m+1}$$

and

$$p(d,m) = \frac{1}{m+1} + \frac{m}{m+1} \cdot p(d-1, m-1).$$

Hence $p(d,n) < O(\frac{d}{n})$. □

Assign to any vertex v in a PHP-tree T a subset $\mathrm{Path}_T(v) \subseteq [n+1] \times [n]$ as follows. If $v_0, \ldots, v_k$ is the path from the root v_0 to $v = v_k$ then

$$\mathrm{Path}_T(v) \ := \ \{(i_0,j_0), \ldots, (i_{k-1}, j_{k-1})\}$$

where at node v_t $(t = 0, \ldots, k-1)$ the tree either queries $i_t \to ?$ and the path follows edge j_t, or it queries $? \to j_t$ and follows edge i_t.

Now observe that if the path in T defined by a sample $\omega \in \Omega^0_{\mathrm{PHP}}$ gets to a node v, then necessarily $\mathrm{Path}_T(v) \subseteq \omega$. In particular, no sample ω ever gets to a node v for which $\mathrm{Path}_T(v)$ is not a partial injective function. In other words, because we let the trees compute only on samples from Ω^0_{PHP}, we could have defined the PHP-trees differently: At node v the tree would not branch according to all is or js but only to those edges (i,j) such that neither i nor j occurs in $\mathrm{Path}_T(v)$.

In particular, if $v = T(\omega)$ is defined then $\mathrm{Path}_T(v)$ is a partial injective map of size at most $\mathrm{dp}(T)$.

A **labeled PHP-tree** (T, ℓ) is a PHP-tree T whose leafs are labeled by elements of $\mathcal{M}_n$. It computes a partial function from Ω^0_{PHP} into $\mathcal{M}_n$: the function maps ω to the label of $T(\omega)$, if $T(\omega)$ is defined.

We will need to talk about predicates on Ω^0_{PHP}, i.e. Boolean functions on Ω^0_{PHP}, being computed by a 0/1-valued PHP-tree. However, the function defined by a PHP-tree is only partially defined and so we need to use a slightly more relaxed notion.

Definition 20.1.4 Let f be a Boolean function with domain Ω^0_{PHP} and (T, ℓ) a PHP-tree labeled by 0 or 1.

Then (T,ℓ) **represents** f iff

$$\ell(T(\omega)) = f(\omega)$$

whenever $T(\omega)$ is defined.

Consider propositional formulas built from variables p_{ij}, $i \in [n+1]$ and $j \in [n]$. A **map-term** is a conjunction of the form

$$p_{i_1 j_1} \wedge \cdots \wedge p_{i_k j_k}$$

where all $i_1,\ldots,i_k$ are different and also all $j_1,\ldots,j_k$ are different. We shall write the map-terms usually without the $\wedge$ sign simply as $p_{i_1 j_1}\ldots p_{i_k j_k}$.

A map-term as above corresponds to a partial injective map

$$a := \{(i_1,j_1),\ldots,(i_k,j_k)\}$$

from $[n+1]$ into $[n]$ in the following sense: an assignment $\omega \in \Omega^0_{\mathrm{PHP}}$ satisfies the map-term iff $a \subseteq \omega$.

A k**-disjunction** is a disjunction of map-terms each of which has size at most k. Such a disjunction is either true or false on any assignment from Ω^0_{PHP} and hence defines a total Boolean function on Ω^0_{PHP}. We will also represent such a function by a $0/1$-valued PHP-tree; this is meant in the precise sense of Definition 20.1.4. In particular, if f is a k-disjunction and (T,ℓ) a $0/1$-valued tree representing f then the tree computes the same value as f whenever it is defined.

Having a $0/1$-valued PHP-tree (T,ℓ) of depth d we define a d-disjunction $D_{(T,\ell)}$ as follows: include a term $p_{i_1 j_1}\ldots p_{i_d j_d}$ for each set $\{(i_1 j_1),\ldots,(i_d,j_d)\}$ that equals to Path_v for some leaf v labeled by 1. The following simple statement will be useful in the later construction of Skolem functions.

Lemma 20.1.5 *(T,ℓ) represents the function defined by $D_{(T,\ell)}$. $D_{(T,\ell)}$ equals* 0 *at those samples where (T,ℓ) is undefined.*

Note that if (T,ℓ) represents f then this disjunction $D_{(T,\ell)}$ may not be equivalent to f.

20.2 Structure $K(F^0_{\mathrm{PHP}}, G^0_{\mathrm{PHP}})$ and open comprehension and open induction

Functions in the family F^0_{PHP} will be defined by the labeled trees and hence will be only partially defined. This is a key feature of the structure. For such partially defined α, β we put

$$\langle\langle \alpha = \beta \rangle\rangle \ := \ \{\omega \in \Omega^0_{\text{PHP}} \mid \text{ both } \alpha(\omega) \text{ and } \beta(\omega) \text{ are defined and equal}\}$$

and as before

$$[\![\alpha = \beta]\!] \ := \ \langle\langle \alpha = \beta \rangle\rangle / \mathcal{I}.$$

Similarly for other relations $R(\alpha_1, \ldots, \alpha_k)$: if any argument α_i is undefined at ω then ω does not get into $\langle\langle R(\alpha_1, \ldots, \alpha_k) \rangle\rangle$ and hence does not count for the truth value $[\![R(\alpha_1, \ldots, \alpha_k)]\!]$. Note that with this definition, for example,

$$\langle\langle \alpha = \alpha \rangle\rangle \neq \Omega^0_{\text{PHP}}.$$

We need to see how this conforms with the earlier set-up where random variables were defined everywhere. We also need to have the equality axioms valid in all structures.

In general this could be a fatal problem. However, elements of F^0_{PHP} will be undefined only on an infinitesimal fraction of Ω^0_{PHP}. Hence we can imagine that the functions are defined on the whole sample space, giving arbitrary values to the samples in the original region of undefinability. This arbitrary extension to total functions does change the $\langle\langle \ldots \rangle\rangle$ values of the atomic sentences as defined above, but only by sets of infinitesimal counting measure. Hence the truth values $[\![\ldots]\!]$ of atomic sentences, and therefore of all sentences, remain unchanged.

In the next definition and in the subsequent text we will therefore talk freely about partially defined random variables.

Now we are ready to define, in a complete analogy with the definition of F_{rud} in Chapter 7, family F^0_{PHP}.

Definition 20.2.1 F^0_{PHP} is the set of all partial functions $\alpha : \Omega^0_{\text{PHP}} \to \mathcal{M}_n$ such that there is a labeled PHP-tree (T, ℓ) satisfying the following:

- $\alpha(\omega)$ is defined iff $T(\omega)$ is defined, for all $\omega \in \Omega^0_{\text{PHP}}$.
- $\alpha(\omega) = \ell(T(\omega))$, for all $\omega \in \Omega^0_{\text{PHP}}$ such that $T(\omega)$ is defined.
- The depth $\text{dp}(T)$ of T satisfies $\text{dp}(T) < n^{1/t}$, for some $t > \mathbf{N}$.

By Lemma 20.1.3 each α is indeed undefined on at most an infinitesimal fraction of Ω^0_{PHP}.

Analogously with the structure $K(F_{\text{rud}}, G_{\text{rud}})$ we shall interpret any m-tuple $\hat{\beta} = (\beta_0, \ldots, \beta_{m-1}) \in \mathcal{M}$ of elements of F^0_{PHP} for an $m \in \mathcal{M}_n$ as defining a function on F^0_{PHP}

$$\hat{\beta}(\alpha) \ : \ \Omega^0_{\text{PHP}} \to \mathcal{M}_n$$

by the condition:

$$\hat{\beta}(\alpha)(\omega) = \begin{cases} \beta_{\alpha(\omega)}(\omega) & \text{if } \alpha(\omega) < m \\ 0 & \text{otherwise.} \end{cases}$$

It is easy to check that any such $\hat{\beta}$ maps F^0_{PHP} into F^0_{PHP}. Thus we can define G^0_{PHP} analogously as a few times before.

Definition 20.2.2 Family G^0_{PHP} consists of all random variables Θ computed by some $\hat{\beta} = (\beta_0, \ldots, \beta_{m-1})$, where $m \in \mathcal{M}_n$ and $\hat{\beta} \in \mathcal{M}$ is a m-tuple of elements of F^0_{PHP}.

The following lemma is obvious; it states properties of $K(F_{\mathrm{rud}}, G_{\mathrm{rud}})$ established in Lemma 7.3.1.

Lemma 20.2.3 *The model $K(F^0_{\mathrm{PHP}}, G^0_{\mathrm{PHP}})$ has the following two properties:*

1. *F^0_{PHP} is*
   (a) L_n-closed, and
   (b) closed under definition by cases by open $L^2_n(F^0_{\mathrm{PHP}}, G^0_{\mathrm{PHP}})$-formulas.
2. *Both F^0_{PHP} and G^0_{PHP} are compact.*

Before the next lemma recall the specific concept of a $0/1$-tree representing a predicate on Ω^0_{PHP} from Definition 20.1.4. The statement is analogous to Lemma 8.2.1 but we shall sketch the proof, because we use the notion of representing a function rather than computing it as in Lemma 8.2.1.

Lemma 20.2.4 *Let $A(y)$ be an open $L^2_n(F^0_{\mathrm{PHP}}, G^0_{\mathrm{PHP}})$-formula with y the only free variable. Then there is $t > \mathbf{N}$ such that for all $i \in \mathcal{M}_n$ the membership in $\langle\langle A(i)\rangle\rangle$ is represented by a PHP-tree with the leafs labeled by* 1 *(for* yes*) or* 0 *(for* no*) of the depth bounded above by $n^{1/t}$. In particular, the membership predicate in $\langle\langle A(i)\rangle\rangle$ is represented by an element of F^0_{PHP}.*

Proof: Proceed by the logical complexity of the formula. Take atomic formula $R(\alpha_1, \ldots, \alpha_k, y)$. The tree β representing $\langle\langle R(\alpha_1, \ldots, \alpha_k, i)\rangle\rangle$ is the tree α_1 to the leafs of which we attach α_2 etc., and at the bottom decide if the relation holds or not. Note that β is defined iff all α_j are, so the sample where β is undefined are also not in $\langle\langle R(\alpha_1, \ldots, \alpha_k, i)\rangle\rangle$.

Conjunction $B(y) \wedge C(y)$ corresponds to composing the trees β_i and γ_i attached to the respective formulas. The negation is treated simply: if we have trees β_i for $B(y)$ then trees β'_i for $\neg B(y)$ are obtained by switching the labels 0 and 1 of the leafs. But note that β and β'_1 are undefined at the same samples and hence this does not correspond to the complement of $\langle\langle B(i)\rangle\rangle$ in $\mathcal{A}$. □

With Lemmas 20.2.3 and 20.2.4 in hand the verification that open comprehension and open induction are valid in $K(F^0_{\mathrm{PHP}}, G^0_{\mathrm{PHP}})$ is done in the same way as for structure $K(F_{\mathrm{rud}}, G_{\mathrm{rud}})$ in Sections 8.2 and 8.3.

Lemma 20.2.5 *Open comprehension and open induction are valid in structure $K(F^0_{\mathrm{PHP}}, G^0_{\mathrm{PHP}})$.*

20.3 The failure of PHP in $K(F^0_{\mathrm{PHP}}, G^0_{\mathrm{PHP}})$

Now we come to the key specific property of the structure.

Lemma 20.3.1

$$[\![\mathrm{PHP}(n, \Delta^0)]\!] = 0_{\mathcal{B}}.$$

Proof: We need to verify that all three disjuncts of PHP(n, Δ^0) get value $0_{\mathcal{B}}$. This is quite straightforward to verify for the second and the third disjuncts. The second disjunct is:

$$\exists v, v' \in [n] \exists u \in [n+1]\ (v \neq v' \wedge \Delta^0(u) = v \wedge \Delta^0(u) = v').$$

We need to verify that for any α, β, β' from F^0_{PHP} the set of samples

$$\langle\!\langle \alpha \in [n+1] \wedge \beta \in [n] \wedge \beta' \in [n] \wedge \beta \neq \beta' \wedge \Delta^0(\alpha) = \beta \wedge \Delta^0(\alpha) = \beta' \rangle\!\rangle$$

has an infinitesimal counting measure. But that is obvious as all samples are partial functions and hence assign to a pigeon at most one hole.

Similarly the third disjunct

$$\exists u, u' \in [n+1] \exists v \in [n]\ (u \neq u' \wedge \Delta^0(u) = v \wedge \Delta^0(u') = v)$$

gets value $0_{\mathcal{B}}$: any sample is a partial injective function, i.e. in any hole sits just one pigeon.

Checking that the first disjunct in PHP(n, Δ^0),

$$\exists u \in [n+1] \forall v \in [n]\ \Delta^0(u) \neq v,$$

gets $0_{\mathcal{B}}$ too is a bit less straightforward. To check this means checking that for any $\alpha \in F^0_{\mathrm{PHP}}$ we have

$$[\![\alpha \in [n+1] \rightarrow \Delta^0(\alpha) \in [n]]\!] = 1_{\mathcal{B}}.$$

Assume that α is determined by a labeled PHP-tree (T, ℓ). The element $\Delta^0(\alpha)$ is computed by (T', ℓ'), a labeled PHP-tree defined as follows:

1. If the label $i = \ell(v)$ of a leaf v of T satisfies $i \in [n+1]$ extend T at v by a depth 1 tree: query $i \to ?$, and label any new leaf corresponding to an edge $j \in [n]$ by j.
2. If $\ell(v) \notin [n+1]$ do nothing.

We need to show that the set

$$\langle\langle \alpha \in [n+1] \wedge \Delta^0(\alpha) \notin [n] \rangle\rangle$$

has an infinitesimal counting measure.

It follows from the definition of Δ^0 that this set is equal to the set of samples ω for which $\Delta^0(\alpha)$ is undefined. But this set has an infinitesimal counting measure by Lemma 20.1.3. □

We leave it to the reader to verify that a similar argument proves that in $K(F^0_{\mathrm{PHP}}, G^0_{\mathrm{PHP}})$ the function Δ^0 is surjective:

$$[\![\forall y \in [n] \exists x \in [n+1]\ \Delta^0(x) = y]\!] \;=\; 1_{\mathcal{B}}.$$

21
Model $K(F_{\mathrm{PHP}}, G_{\mathrm{PHP}})$ of V_1^0

We are going to modify the structure $K(F^0_{\mathrm{PHP}}, G^0_{\mathrm{PHP}})$ from the previous chapter to a new structure $K(F_{\mathrm{PHP}}, G_{\mathrm{PHP}})$. This new structure will be a model of V_1^0.

The construction of $K(F_{\mathrm{PHP}}, G_{\mathrm{PHP}})$ is quite analogous to the construction of $K(F_{\mathrm{tree}}, G_{\mathrm{tree}})$ from $K(F_{\mathrm{rud}}, G_{\mathrm{rud}})$: it is based on what could be called deep PHP-trees. And, as before, the main purpose of the more complicated definition is to get bounded quantifier elimination. For this we will utilize a switching lemma from Krajíček *et al.* [79] and Pitassi *et al.* [88], and we start by describing it.

21.1 The PHP switching lemma

In this section we shall recall the switching lemma involving PHP-trees, developed in [79, 88] for the purpose of improving Ajtai's original, mere super-polynomial, lower bound [2] in Theorem 19.1.2. The lemma is more involved than the basic example of Hastad's switching lemma (Theorem 10.1.2) as it concerns truth assignments where there is a great deal of dependence between values of propositional variables (in Theorem 10.1.2 these are completely independent).

We need first to define the space of partial truth assignments (restrictions) used in the switching lemma. Assume $D \subseteq [n+1]$ and $R \subseteq [n]$ are two sets of equal size. Denote by $\mathrm{Res}(D,R)$ the set of all partial injective maps from $[n+1]$ into $[n]$ with domain D and range R. If $|D| = m$ then clearly there are $m!$ such maps.

Now we are ready to state the switching lemma in the form in which we shall use it.

Theorem 21.1.1 (Krajíček *et al.* [79], Pitassi *et al.* [88]) *Let $1 \leq m \leq n$ be arbitrary. Let $1 \leq a \leq b$. Assume a Boolean function f on Ω^0_{PHP} is defined by an a-disjunction.*

Pick a subset $D \subseteq [n+1]$ and a subset $R \subseteq [n]$ both of size m, uniformly at random, and then pick a restriction $\rho \in \mathrm{Res}(D,R)$, also uniformly at random.

Then the probability that the restricted function $f \downarrow \rho$ is not on the set of truth assignments $\{\omega \in \Omega^0_{PHP} \mid \rho \subseteq \omega\}$ representable by a PHP-tree of depth at most b is bounded above by

$$\left[\frac{(n+1-m)^4 a^3)}{m}\right]^{b/2}.$$

We will be picking m of the form $n - n^\delta$, for $\delta > 0$ a standard rational. But note that contrary to Theorem 10.1.2 we cannot pick δ arbitrarily close to 1, even for a of the form n raised to an infinitesimal; we need $\delta < \frac{1}{4}$ (we will use $\frac{1}{5}$).

21.2 Structure $K(F_{PHP}, G_{PHP})$

The definition of $K(F_{PHP}, G_{PHP})$ is analogous to that of $K(F_{tree}, G_{tree})$ but there are differences. This is due to the remark after Theorem 21.1.1 and to the fact that it is not so straightforward to represent the set of restrictions $\mathrm{Res}(D,R)$ by a tree. The formalism we define bellow could have been used in the definition of $K(F_{tree}, G_{tree})$ as well.

Definition 21.2.1 For $k \geq 0$ define a **level k chain**, or briefly a k-chain, as follows. The 0-chain is the triple $(\emptyset, \emptyset, \emptyset)$.

For $k \geq 1$, a k-chain T is a $(k+1)$-tuple

$$(D_i, R_i, \rho_i)\ _{0 \leq i \leq k}$$

such that:

- $D_0 = R_0 = \rho_o = \emptyset$
- $D_1 \subseteq [n+1]$ and $R_1 \subseteq [n]$ are sets of size $n - n^{1/5}$ (rounded down), and $\rho_1 \in \mathrm{Res}(D_1, R_1)$
- for $i = 2, \ldots, k$, $D_i \subseteq [n+1] \setminus \bigcup_{0 \leq j < i} D_j$ and $R_i \subseteq [n] \setminus \bigcup_{0 \leq j < i} R_j$ are sets of size $n^{(1/5)^{i-1}} - n^{(1/5)^i}$
- for $i = 1, \ldots, k$, $\rho_i \in \mathrm{Res}(D_i, R_i)$.

We make the provision that if the size of ρ_k should drop below 10 then D_k, R_k and ρ_k are of maximal possible sizes, i.e. $R_k = [n]$, D_k is $[n+1]$ without one pigeon and $|\rho_k| = n$.

Let $\rho(T) := \bigcup_{0 \le j \le k} \rho_j$. The support of T is the pair of sets $(\bigcup_{0 \le j \le k} D_j, \bigcup_{0 \le j \le k} R_j)$.

A partial ordering $\preceq$ on these chains is defined as follows: for a k-chain T and an ℓ-chain S it holds that $T \preceq S$ iff T is an initial part of S.

Chain_k is the set of all k-chains.

Note that if T and S are both k-chains then $T \preceq S$ iff $T = S$. Also note that the complement of the support of a k-chain contains $n^{5^{-k}} + 1$ pigeons and $n^{5^{-k}}$ holes.

Denote by h_0 the minimal number such that the support of h_0-chains contains all pigeons except one, and all holes. That is, the restriction $\rho(T)$ defined by an h_0-chain T has size n (and, in particular, is in Ω^0_{PHP}). Constant h_0 has the size around $O(\log(\log(n)))$.

Definition 21.2.2 The sample space Ω_{PHP} is the set Chain_{h_0}.

Note that any $\omega \in \Omega^0_{\mathrm{PHP}}$ is equal to $\rho(T)$ for exponentially many h_0-chains T. In a sense the samples in Ω_{PHP} are those in Ω^0_{PHP} but augmented with a particular history of how they were obtained by a succession of restrictions (of specific forms).

The definition of the family of random variables F_{PHP} is now analogous to Definition 10.2.4.

Definition 21.2.3 The family F_{PHP} consists of all functions $\alpha \in \mathcal{M}$,

$$\alpha : \Omega_{\mathrm{PHP}} \to \mathcal{M}_n,$$

for which there exists $k \in \mathbf{N}$, $t > \mathbf{N}$ and a tuple $(S_Z)_{Z \in \mathrm{Chain}_k} \in \mathcal{M}$ of PHP-trees such that

- Each S_Z is a PHP-tree.
- The depth of all S_Z in the tuple is bounded by $n^{1/t}$.
- The value of α on sample $T \in \Omega_{\mathrm{PHP}}$ is computed by S_Z at $\omega := \rho(T)$, for $Z \preceq T$.

The parameter k is called the level of α.

Note that for each $Z \in \mathrm{Chain}_k$, $k \in \mathbf{N}$, Lemma 20.1.3 implies that S_Z fails to be defined for an infinitesimal fraction $O(n^{1/t}/n^{5^{-k}})$ of $\{T \in \Omega_{\mathrm{PHP}} \mid Z \preceq T\}$. Hence the region of undefinability of an element of F_{PHP} has an infinitesimal counting measure.

Note also that, without loss of generality, we may assume that the PHP-tree S_Z queries only pigeons/holes outside the support of Z (see the remark after Lemma 20.1.3).

We make the usual simple technical observation, with a proof analogous to the proof of Lemma 10.2.5.

Lemma 21.2.4 *Let $m \in \mathcal{M}_n$ and let $\hat{\beta} = (\beta_0, \ldots, \beta_{m-1}) \in \mathcal{M}$ be an m-tuple of elements of F_{PHP}. For any $\alpha \in F_{\mathrm{PHP}}$ define function $\hat{\beta}(\alpha) : \Omega_{\mathrm{PHP}} \to \mathcal{M}_n$ by*

$$\hat{\beta}(\alpha)(T) = \begin{cases} \beta_{\alpha(T)}(T) & \text{if } \alpha(T) < m \\ 0 & \text{otherwise.} \end{cases}$$

Then this function is in F_{PHP} too.

With the lemma in hand we can define the family G_{PHP} in the same way as we defined G_{tree} or G^0_{PHP} earlier.

Definition 21.2.5 Let $m \in \mathcal{M}_n$ and let $\hat{\beta} = (\beta_0, \ldots, \beta_{m-1}) \in \mathcal{M}$ be an m-tuple of elements of F_{PHP}.

For any $\alpha \in F_{\mathrm{PHP}}$ define $\hat{\beta}(\alpha) : \Omega_{\mathrm{PHP}} \to \mathcal{M}_n$ by

$$\hat{\beta}(\alpha)(T) = \begin{cases} \beta_{\alpha(T)}(T) & \text{if } \alpha(T) < m \\ 0 & \text{otherwise.} \end{cases}$$

Family G_{PHP} consists of all random variables Θ computed by some $\hat{\beta} = (\beta_0, \ldots, \beta_{m-1})$, where $m \in \mathcal{M}_n$ and $\hat{\beta} \in \mathcal{M}$ is a m-tuple of elements of F_{PHP}.

We conclude this section with a technical lemma useful for open comprehension and open induction, and later for the bounded quantifier elimination. It is again quite analogous to Lemma 10.2.7, and to Lemmas 20.2.3 and 20.2.4.

Lemma 21.2.6 **(1)** *Let $A(x)$ be an open $L^2_n(F_{\mathrm{PHP}}, G_{\mathrm{PHP}})$-formula with the only free variable x. Then there are $k \in \mathbf{N}$ and $t > \mathbf{N}$ such that for all $i \in \mathcal{M}_n$ the membership in $\langle\langle A(i) \rangle\rangle$ is represented by a $0/1$-valued element of F_{PHP} having level k and depth bounded above by $n^{1/t}$.*
(2) *Let $s > \mathbf{N}$ and $w := n^{1/s}$. Assume $(\alpha_1, \ldots, \alpha_w) \in \mathcal{M}$ is an w-tuple of elements of F_{PHP}. Then there is $\beta \in F_{\mathrm{PHP}}$ such that*

- *β is defined at a sample iff all α_i are defined*
- *β computes simultaneously all α_i in the tuple (i.e. it computes the w-tuple of the values).*

(3) *Family F_{PHP} is closed under definition by cases by open $L^2_n(F_{\mathrm{PHP}}, G_{\mathrm{PHP}})$-formulas.*

21.3 Open comprehension, open induction and failure of PHP

Recall from Chapter 7 that comprehension for open formula $A(x)$ on $[0,m)$ is witnessed by an m-tuple $\hat{\beta} = (\beta_0, \ldots, \beta_{m-1})$, where β_i represents the predicate $\langle\langle A(i) \rangle\rangle$. Similarly, induction for $A(x)$ up to m is witnessed by a random variable computed by a tree that is built from trees for β_i in a form simulating the binary search. Hence Lemma 21.2.6 (part (1)) gives us what we need to carry this proof in $K(F_{\text{PHP}}, G_{\text{PHP}})$ as well.

Lemma 21.3.1 *Open comprehension and open induction are valid in structure* $K(F_{\text{PHP}}, G_{\text{PHP}})$.

We also need to verify that PHP fails in $K(F_{\text{PHP}}, G_{\text{PHP}})$. But we have to first define a suitable map from G_{PHP}.

Definition 21.3.2 Element $\Delta \in G_{\text{PHP}}$ is defined as follows: for a sample $T \in \Omega_{\text{PHP}}$ put

$$\Delta_T := \rho(T).$$

That is, Δ is determined by the $(n+1)$-tuple $(\beta_i)_{i \in [n+1]}$ where β_i is a depth 1 tree querying $i \to ?$ at the root and labeling the leaf on the edge corresponding to $j \in [n]$ by j itself.

Note that $\Delta_T = \Delta^0_\omega$ for $\omega = \rho(T)$, so the definition of Δ just sheds the extra chain structure on T but is otherwise as in Δ^0.

Lemma 21.3.3

$$[\![\text{PHP}(n, \Delta)]\!] = 0_{\mathcal{B}}.$$

Proof: The proof is essentially identical to that of Lemma 20.3.1; we just need to find our way in the slightly more complicated structure $K(F_{\text{PHP}}, G_{\text{PHP}})$. Let us treat only the first disjunct of PHP(n,R), the hardest case in the proof of Lemma 20.3.1.

We want to show that

$$[\![\alpha \in [n+1] \to \Delta(\alpha) \in [n]]\!] = 1_{\mathcal{B}},$$

i.e. that the set

$$(*) \qquad \langle\langle \alpha \in [n+1] \to \Delta(\alpha) \in [n] \rangle\rangle$$

differs from Ω_{PHP} by a set of an infinitesimal counting measure. Function α is undefined on at most an infinitesimal fraction of the sample space. If it is defined at a sample T and $\alpha(T) \notin [n+1]$, T gets into set $(*)$. If $\alpha(T)$ is defined and equal to some $i \in [n+1]$ then either $\rho(T)(i) = j \in [n]$ and T gets into $(*)$, or $\rho(T)$ is undefined. The latter happens with infinitesimal probability $(n^{5^{-k}}+1)^{-1}$, if α is of level k. □

21.4 Bounded quantifier elimination

What remains is to construct suitable Skolem functions. The construction is analogous to that of Skolem functions in $K(F_{\text{tree}}, G_{\text{tree}})$ but there some differences in the analysis of what is going on. This is discussed in detail later, in Section 22.5, and some readers may prefer to read that exposition now.

Theorem 21.4.1 *Structure $K(F_{\text{PHP}}, G_{\text{PHP}})$ is Skolem closed.*

Proof: We need to show that for any open first-order $L_n^2(F_{\text{PHP}}, G_{\text{PHP}})$-formula $B(x,y)$ and for any $m \in \mathcal{M}_n$ there is a $\Theta \in G_{\text{PHP}}$ that is a Skolem function for $\exists y B(x,y)$ on m.

The strategy of the proof is the same as that of the proofs of Theorems 11.1.1 and 15.1.1. We want to find an m-tuple $\hat{\beta} = (\beta_0, \ldots, \beta_{m-1}) \in \mathcal{M}$ of elements of F_{PHP} such that each of the sets

$$E_i := \langle\langle \exists y < m B(i,y) \rangle\rangle \setminus \langle\langle \beta_i < m \wedge B(i, \beta_i) \rangle\rangle,$$

$i = 0, \ldots, m-1$, has an infinitesimal counting measure in Ω_{PHP} and such that, in fact, the union of these sets has an infinitesimal counting measure too. Our first claim is identical to that in the proofs of Theorems 11.1.1 and 15.1.1 and holds for exactly the same reasons (we repeat the argument for the sake of completeness).

Claim 1: *Assume that the counting measure (in Ω_{PHP}) of $\bigcup_{i<m} E_i \in \mathcal{A}$ is infinitesimal. Then $\Theta \in G_{\text{PHP}}$ computed by $\hat{\beta}$ is a Skolem function for $\exists y < m B(x,y)$ on m.*

To prove the claim assume for the sake of a contradiction that

$$[\![\exists y < m\, B(\alpha, y)]\!] \; > \; [\![\Theta(\alpha) < m \wedge B(\alpha, \Theta(\alpha))]\!]$$

for some α. Applying part (3) of Lemma 21.2.6 and Lemma 3.3.2 we get γ such that

$$[\![\gamma < m \wedge B(\alpha, \gamma)]\!] \; > \; [\![\Theta(\alpha) < m \wedge B(\alpha, \Theta(\alpha))]\!].$$

Therefore

$$U \; := \; \langle\langle \gamma < m \wedge B(\alpha, \gamma) \rangle\rangle \setminus \langle\langle \Theta(\alpha) < m \wedge B(\alpha, \Theta(\alpha)) \rangle\rangle$$

has a non-infinitesimal counting measure in Ω_{PHP}. But clearly

$$U \subseteq \bigcup_{i<m} \left(\{\omega \in \Omega_{\mathrm{PHP}} | \, \alpha(\omega) = i\}\right) \cap E_i \subseteq \bigcup_{i<m} E_i,$$

which is a contradiction. This proves the claim.

Hence we want to construct $\beta_i \in F_{\mathrm{PHP}}$ satisfying the hypothesis of Claim 1. We have:

$$\langle\langle \exists y < mB(i,y) \rangle\rangle \; = \; \bigcup_{j<m} \langle\langle B(i,j) \rangle\rangle.$$

By part (1) of Lemma 21.2.6 the membership in sets $\langle\langle B(i,j) \rangle\rangle$ are represented by 0/1-valued functions $\xi_{i,j}$ from F_{PHP} such that all $\xi_{i,j}$ have the same standard level k and depth bounded above by $n^{1/t}$, for some common non-standard t .

Claim 2: *For any $i < m$, $0 \le u < v < m$ and $s > \mathbf{N}$ there are level $k+1$ elements $\kappa_{i,u,v} = (R_W^{i,u,v})_{W \in \mathrm{Chain}_{k+1}}$ of F_{PHP} of depth at most $n^{\frac{1}{s}}$ such that:*

- *The probability that, for a random $W \in \mathrm{Chain}_{k+1}$, tree $R_W^{i,u,v}$ does not represent for $W \preceq T$ the predicate*

$$T \in \langle\langle \exists u \le y \le v \; B(i,y) \rangle\rangle$$

is bounded above by

$$2^{-n^{1/s}}.$$

We shall use Theorem 21.1.1 to prove this key claim. We fix parameters i,u,v,s satisfying the restrictions of the claim (and do not otherwise reflect these parameters in the forthcoming notation). Also put $a := n^{1/t}$and $b := n^{1/s}$.

Let $(S_Z^j)_Z$, for $u \le j \le v$ and $Z \in \mathrm{Chain}_k$, be the trees of depth at most a computing $\xi_{i,j}$. Let D_Z^j be the disjunction of all map-terms corresponding to paths in S_Z^j ending with label 1 (for yes). Note that each D_Z^j is an a-DNF formula.

The element $\kappa_{i,u,v} \in F_{\mathrm{PHP}}$ will be computed by trees $(R_W)_{W \in \mathrm{Chain}_{k+1}}$, the trees having depth at most b. We are now going to argue, using Theorem 21.1.1, that there exist such trees R_W such that the element $\kappa_{i,u,v}$ defined by them satisfies the requirement of the claim.

Fix $Z \in \mathrm{Chain}_k$. Let D_Z be the disjunction $\bigvee_{u \le j \le v} D_Z^j$. It is an a-DNF formula that has variables corresponding to the complement of the support of Z. That is, there are $n^{5^{-k}}+1$ possible pigeons and $n^{5^{-k}}$ possible holes. By its definition the formula defines correctly for all T such that $Z \preceq T$, with an infinitesimal probability of an error, the predicate:

$$\left(T \in \bigcup_{u \le j \le v} \langle\langle B(i,j) \rangle\rangle \right).$$

Take a random $(k+1)$-chain W, $Z \preceq W$. By Theorem 21.1.1 there exists a PHP-tree R_W such that

- R_W represents $D_Z \downarrow \rho(W)$, and
- R_W has the depth at most b,

with the probability of failing to exist for a random W at most

$$\left[\frac{(n^{5^{-(k+1)}}+1)^4 a^3}{n^{5^{-k}} - n^{5^{-(k+1)}}} \right]^{b/2} < 2^{-n^{1/s}}.$$

If there exists a suitable R_W, we conclude the argument by the following reasoning (discussed further in Section 22.5). Namely, we argue that, for any α, $\kappa_{i,u,v}$ represents the predicate $\langle\langle u \le \alpha \le v \wedge B(i,\alpha) \rangle\rangle$.

Fixing Z and ignoring the $u \le \alpha \le v$ part, the predicate

$$\langle\langle B(i,\alpha) \rangle\rangle \cap \{T \mid Z \preceq T\}$$

is represented by a PHP-tree $\Gamma(\alpha)$, where Γ is the map defined by the m-tuple $(S_Z^j)_{j<m}$. So it suffices to verify that, for $Z \preceq T$

$$\Gamma(\alpha)(T) = 1 \rightarrow \kappa_{i,u,v}(T) = 1$$

if they are both defined. If $\Gamma(\alpha)(T) = 1$ then $\alpha(T)$ is defined and equal to some $j < m$, and $S_Z^j(T)$ is defined and equal to 1. So $D_Z^j(T) = 1$ and $D_Z(T) = 1$, and $\kappa_{i,u,v}(T) = 1$ too.

But, by the estimate above, suitable trees R_W do exist for all but a fraction of at most $2^{-n^{1/s}}$ chains W. This proves the claim.

To define elements β_i finding a j such that $T \in \langle\langle B(i,j) \rangle\rangle$ if it exists (with only an infinitesimal probability of failing) we simulate the binary search using functions $\kappa_{i,u,v}$. This requires us to compute $\log(m)$ of such functions on a path in the search, and the whole binary search can be performed by an element β_i of F_{PHP} (as in Lemma 10.2.7, part 2). The depth of the trees computing β_i is bounded above by $\log(m) \cdot b$ and that obeys the restrictions for a $(k+1)$-level element of F_{PHP}.

By Claim 2 each $\kappa_{i,u,v}$ makes an error with probability less than $2^{-n^{1/s}}$. There are at most m such pairs $u \leq v$ used in a binary search and there are m different instances i. Hence the counting measure of the set of samples where we can make an error is less than

$$2^{-n^{1/s}} \cdot m^2.$$

By taking s non-standard but small enough this quantity can be made infinitesimal. □

The theorem yields, via Theorem 9.2.3, the bounded quantifier elimination.

Corollary 21.4.2 *The model $K(F_{\mathrm{PHP}}, G_{\mathrm{PHP}})$ admits bounded quantifier elimination.*

21.5 PHP lower bound for F_d: a summary

Proof of Theorem 19.1.2

By Corollary 21.4.2 and Lemma 21.3.1 the structure $K(F_{\mathrm{PHP}}, G_{\mathrm{PHP}})$ is a model of V_1^0. By Lemma 21.3.3 the PHP principle fails in $K(F_{\mathrm{PHP}}, G_{\mathrm{PHP}})$; its truth value is $0_{\mathcal{B}}$. This proves the first part of the theorem.

The second part, the lower bound for F_d-proofs of PHP_n, follows from the first part by Theorem 19.2.1. □

22
Algebraic PHP model?

This chapter is devoted to the problem of a construction of an L_n-closed model of the theory $Q_2V_1^0$ in which PHP would fail. The existence of such a structure would imply via Theorem 19.2.1 the affirmative solution of the following problem.

Problem 22.0.1 *Is it true that for every $d \geq 3$ there is an $\epsilon > 0$ such that proofs of the tautologies* PHP_k *for $k \geq 1$ in a constant-depth Frege system $F_d(\oplus)$ in the DeMorgan language with the parity connective $\oplus$ must have size at least 2^{k^ϵ}?*

In particular, is it true that the theory $Q_2V_1^0$, even augmented by any function of a subexponential growth, does not prove $\forall x \mathrm{PHP}(x,R)$?

The problem of proving a lower bound for $F_d(\oplus)$ has been open for quite some time. Neither Ajtai's method [2], so successful for F_d, nor its later improvements by Krajíček *et al.* [79] and Pitassi *et al.* [88] (or any other method for that matter) were able to tackle the problem. On the other hand, before Ajtai's paper, Razborov [92] had already invented a simple method for proving lower bounds for constant-depth circuits in the DeMorgan language with the parity connective (this was subsequently simplified and generalized by Smolensky [101]). Thus, immediately after Ajtai's paper, researchers attempted to modify the Razborov–Smolensky method to a proof complexity setting in order to prove lower bounds for $F_d(\oplus)$. The problem is often branded as the hardest problem (in proof complexity) among those deemed doable (however, it was repeatedly described as doable over the span of more than twenty years).

No lower bounds for the proof system are known but there are several relevant results nevertheless. In particular, Krajíček [58] proves an exponential lower bound for a subsystem of the proof system that extends both constant-depth Frege systems and polynomial calculus, Maciel and Pitassi [81] give a quasi-polynomial simulation of the proof system by a depth 3 subsystem that can also use a threshold connective, and Impagliazzo and Segerlind [47] prove that constant-depth Frege systems with counting axioms modulo a prime p do not polynomially simulate polynomial calculus over $\mathbf{F}_p$. Additional bibliographical information related to the problem can be found in these papers. Constant-depth Frege systems and various algebraic proof systems are also surveyed in Buss *et al.* [18].

In earlier chapters we have developed a uniform approach to the construction of the tree model, the algebraic model and the PHP-tree model, and one may hope that it would give us a hint how to solve the 'equation'

$$\frac{K(F_{\mathrm{PHP}}, G_{\mathrm{PHP}})}{K(F_{\mathrm{tree}}, G_{\mathrm{tree}})} = \frac{?}{K(F_{\mathrm{alg}}, G_{\mathrm{alg}})}$$

and construct a model needed for a solution of the problem.

One may be under the impression (as this author originally was) that solving this equation should be simple and straightforward, and that it will suffice to go where the flow of the method will lead us. In retrospect this appears distinctively over-optimistic.

We have not found the right pair of a sample space and a family of random variables that would define the model we want. But even through failed attempts one learns a few things. Perhaps the most relevant is that, rather than look at the problem as the equation above, it seems more useful to consider the diagram

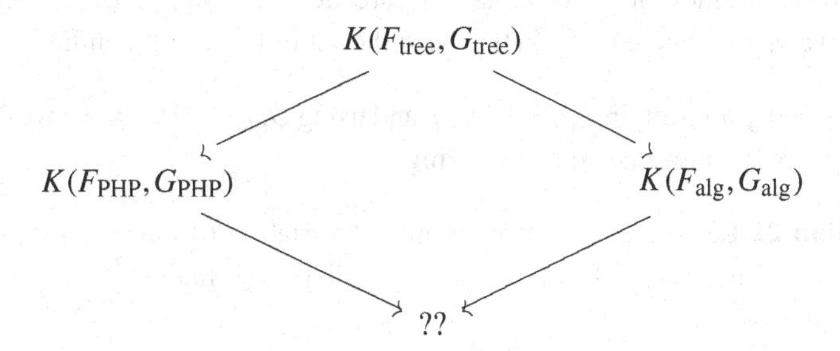

and think how to complete it.

It became clear that the key issue is to come up with the right notion of an 'algebraic PHP-tree'. Such trees will define random variables analogously to trees in earlier structures. However, these random variables have to be only partially defined. This appears to be a relevant insight and we discuss it in detail in Section 22.5.

Before we get to that we shall review in earlier sections some material from proof complexity of algebraic proof systems that may be relevant to the eventual construction of the required structure.

22.1 Algebraic formulation of PHP and relevant rings

Definition 22.1.1 Let $\mathbf{F}_2[x_{ij}]$, with $i \in [n+1]$ and $j \in [n]$, be the ring of polynomials over $\mathbf{F}_2$ in the indicated variables x_{ij}.

S is the ring $\mathbf{F}_2[x_{ij}]$ factored by the ideal generated by all polynomials $x_{ij}^2 - x_{ij}$. In particular, S is an $\mathbf{F}_2$-vector space whose basis is formed by multi-linear monomials.

We say that a polynomial has **low degree** if the degree is bounded above by some term of the form $n^{1/t}$, for a non-standard $t > \mathbf{N}$.

We shall denote by S^{low} the family of polynomials from S of low degree. (Family S^{low} is a subset of $\mathcal{M}$ but not a definable one.)

We shall sometimes abuse the terminology and also call a number from $\mathcal{M}_n$ **low** if it is bounded above by some term of the form $n^{1/t}$, for a non-standard $t > \mathbf{N}$.

Definition 22.1.2 (Beame *et al.* [8]) The $(\neg\text{PHP}_n)$-system is the system of polynomial equations over S consisting of equations:

- $x_{i_1 j} \cdot x_{i_2 j} = 0$, for each $i_1 \neq i_2 \in [n+1]$ and $j \in [n]$
- $x_{ij_1} \cdot x_{ij_2} = 0$, for each $i \in [n+1]$ and $j_1 \neq j_2 \in [n]$
- $1 - \sum_{j \in n} x_{ij} = 0$, for each $i \in [n+1]$.

The left-hand sides of these equations are denoted $Q_{i_1,i_2;j}$, $Q_{i;j_1,j_2}$ and Q_i respectively. We denote by $(\neg\text{PHP}_n)$ also the set of these polynomials.

Note that by multiplying $Q_i = 0$ by x_{ij} and using $Q_{i;j,j'} = 0$ for $j' \neq j$ we derive $x_{ij} - x_{ij}^2 = 0$. We introduce one more ring.

Definition 22.1.3 If $\bar{r} = (r_\ell)_\ell$ are additional variables (different from all x_{ij}) then $\widetilde{S[\bar{r}]}$ is the ring $\mathbf{F}_2[\bar{x}, \bar{r}]$ factored by all polynomials $x_{ij}^2 - x_{ij}$ and all polynomials $r_\ell^2 - r_\ell$.

We shall use the same notation $\widetilde{S[\bar{r}]}$ for rings possibly differing in tuples of $\bar{r}$; this is always clear from the context. Note that $\widetilde{S[\bar{r}]}$ is not $S[\bar{r}]$; it is its quotient by polynomials $r_\ell^2 - r_\ell$ (hence the notation with $\tilde{}$). Every element of $\widetilde{S[\bar{r}]}$ is represented by a canonical multi-linear polynomial over $\mathbf{F}_2$ and hence has a well-defined degree.

We would like to take for the sample space the set of maximal ideals in S (or at least in S^{low}) in $\mathcal{M}$ containing $(\neg\text{PHP}_n)$ and not containing 1. But as $(\neg\text{PHP}_n)$ is unsolvable in $\mathcal{M}$ no such ideal exists.

22.2 Nullstellensatz proof system NS and designs

The following definition makes sense over any commutative ring but we shall specialize (here and in later definitions) to $\widetilde{S[\bar{r}]}$ only.

Definition 22.2.1 (Beame *et al.* [8]) Let $\mathcal{F} = \{f_i \mid i = 1,\ldots,k\}$ be a set of polynomials from $\widetilde{S[\bar{r}]}$ and $f_0 \in \widetilde{S[\bar{r}]}$. A **Nullstellensatz proof** (NS-proof) of f_0 from $\mathcal{F}$ is a k-tuple of polynomials g_i, $i = 1,\ldots,k$, such that

$$g_1 \cdot f_1 + \cdots + g_k \cdot f_k = f_0$$

holds in $\widetilde{S[\bar{r}]}$. The degree of the refutation is the maximum of $\deg(g_i) + \deg(f_i)$, $i \leq k$.

An NS-proof of 1 is called an **NS-refutation** of $\mathcal{F}$.

The notion of design was invented by P. Pudlák[1] for the special case of counting principles and used first in Beame *et al.* [8]. A general definition of designs has been studied by Buss [16]. We specialize to ring S only (and, as in the whole book, to $\mathbf{F}_2$).

Definition 22.2.2 Let $t \geq 0$ be a parameter. Let $\mathcal{F}$ be a set of polynomials from S of degree $\leq t$. A degree t design for $\mathcal{F}$ is a map

$$D : S \to \mathbf{F}_2$$

satisfying the following three conditions:

- $D(0) = 0$ and $D(1) = 1$
- D is a linear map
- $D(g \cdot f) = 0$ for any $f \in \mathcal{F}$ and $g \in S$ such that $\deg(f) + \deg(g) \leq t$.

The raison d'etre for the definition is the following simple theorem (see Beame *et al.* [8, Section 5], Buss *et al.* [18, Section 4]) or Buss [16, Theorem 3]). It follows by applying a design to both sides of a potential NS-refutation, yielding contradiction $0 = 1$.

Theorem 22.2.3 *Let $t \geq 0$ be a parameter. Let $\mathcal{F}$ be a set of polynomials from S of degree $\leq t$.*

If there is a degree t design for $\mathcal{F}$ then $\mathcal{F}$ does not have a degree $\leq t$ Nullstellensatz refutation in S.

In fact, the existence of a degree t design in the theorem is not only sufficient but also necessary; this follows by the duality of linear programming as shown

[1] In the electronic Prague–San Diego seminar that ran in the mid 1990s.

by Buss [14]. Theorem 22.2.3 was used in the proof of the next statement (the original result is for the so-called modular counting principles; the PHP lower bound follows via known and simple reductions).

Theorem 22.2.4 (Beame *et al.* [8], Buss *et al.* [18]) *Any Nullstellensatz refutation of* $(\neg \mathrm{PHP}_n)$ *over* $\mathbf{F}_2$ *must have the degree at least* $n^{\Omega(1)}$.

22.3 A reduction of $F_d(\oplus)$ to NS with extension polynomials

We can represent constants 0 and 1 from S by propositional constants 0 and 1, and monomials in variables x_{ij} by conjunctions of variables occurring in them. If

$$f = \sum_{u=1,\ldots,v} \Pi_{(i,j)\in J_u} x_{ij}$$

is a degree d polynomial from S then the statement $f = 0$ is expressible by a depth 2 propositional formula

$$\neg \bigoplus \left(\bigwedge_{(i,j)\in J_1} x_{ij}, \ldots, \bigwedge_{(i,j)\in J_v} x_{ij} \right).$$

The size of the formula is bounded by the number of monomials of degree d, i.e. by $n^{O(d)}$. In particular, any degree d NS-refutation of a set of polynomials f_i from S can be turned into a constant depth $F(\oplus)$-proof of $\bigvee_i \neg(f_i = 0)$ of size $n^{O(d)}$.

Buss *et al.* [18] found a way how to augment Nullstellensatz so that also the opposite implication holds. Here the ring $\widetilde{S[\bar{r}]}$ enters the picture.

Definition 22.3.1 (Buss *et al.* [18]) Let $\bar{g} = g_1, \ldots, g_m$ be polynomials from $\widetilde{S[\bar{r}]}$ and let $h \geq 1$ be a fixed number. Take new variables r_{iu}, $i = 1, \ldots, m$ and $u = 1, \ldots, h$ not occurring in any of g_j.

The polynomials $E_{i,\bar{g}}$, given by

$$g_i \cdot \Pi_{u \leq h} \left(1 - \sum_{j \leq m} r_{ju} g_j \right),$$

one for each $i \leq m$, are called **extension polynomials** of accuracy h corresponding to $\bar{g}$. The variables r_{iu} are called extension variables corresponding to $\bar{g}$.

To understand the meaning of the extension polynomials consider the polynomial

$$f(\bar{x},\bar{r}) := \Pi_{u \le h}\left(1 - \sum_{j \le m} r_{ju} g_j\right).$$

If all $g_i = 0$ then $f(\bar{x},\bar{r}) = 1$. On the other hand, if $f(\bar{x},\bar{r}) = 1$ then – assuming that all extension polynomials corresponding to $\bar{g}$ are zero – necessarily all $g_i = 0$. In this way the polynomial $1 - f(\bar{x},\bar{r})$, which is of degree at most $h \cdot (1 + \max_i \deg(g_i))$, represents the disjunction $\bigvee_i (g_i = 1)$. If we represented this disjunction straightforwardly by $1 - \Pi_i(1 - g_i)$ we would get a polynomial of a possibly much bigger degree $\sum_i \deg(g_i)$.

Definition 22.3.2 (Buss *et al.* [18]) Let $\mathcal{E}$ be a set of extension polynomials from $\widetilde{S[\bar{r}]}$. We call the set $\mathcal{E}$ **leveled** if the following conditions hold:

1. All variables occurring in $\mathcal{E}$ are either $\bar{x}$ or extension variables of some polynomial in $\mathcal{E}$.
2. If $\mathcal{E}$ contains some extension polynomial, $E_{i,g_1,\ldots,g_m}$, then $\mathcal{E}$ must contain all companion extension polynomials $E_{1,\bar{g}}, E_{2,\bar{g}}, \ldots, E_{m,\bar{g}}$.
3. $\mathcal{E}$ can be decomposed into levels $\mathcal{E} = \mathcal{E}_1 \dot{\cup} \ldots \dot{\cup} \mathcal{E}_\ell$ in such a way that for any polynomial $E_{i,g_1,\ldots,g_m} \in \mathcal{E}_j$ its companion polynomials $E_{1,\bar{g}}, \ldots, E_{m,\bar{g}}$ also belong to $\mathcal{E}_j$, and extension variables corresponding to $\bar{g}$ do not occur in any other polynomial from $\mathcal{E}_1 \dot{\cup} \ldots \dot{\cup} \mathcal{E}_j$.

The minimal ℓ for which such a decomposition is possible is called the **depth** of $\mathcal{E}$.

Now we state the reduction of $F_d(\oplus)$ to NS augmented by the extension polynomials. We consider only the case of $(\neg\mathrm{PHP}_n)$ but the reduction applies to a general system of polynomial equations.

Theorem 22.3.3 (Buss *et al.* [18, Theorem 6.7(1)]) *For every $d \ge 3$ there is a constant $c \ge 1$ such that for any $n \ge 2$ and $h \ge 1$ the following holds.*

If there exists an $F_d(\oplus)$-proof of PHP$_n$ *with s inferences then there exists a levelled set $\mathcal{E}$ of extension polynomials with accuracy h such that $|\mathcal{E}| \le s^c$, the depth of $\mathcal{E}$ is at most $d + c$, and $(\neg\mathrm{PHP}_n) \cup \mathcal{E}$ has a Nullstellensatz refutation in $\widetilde{S[\bar{r}]}$ (where $\bar{r}$ are all extension variables) of degree at most* $(\log s)(h+1)^c$.

The theorem is proved by an elementary but lengthy proof-theoretic argument which we shall not repeat here. The idea is to translate the formulas occurring in a proof into low-degree polynomials as they are introduced in the proof, and transform the proof step by step into a Nullstellensatz refutation. Whenever a big disjunction (or conjunction) should be translated, rather than using the straightforward translation one introduces suitable extension polynomials and uses the low-degree translation outlined before Definition 22.3.2. The resulting set of extension polynomials is levelled because the extension polynomials introduced when translating a depth ℓ formula will be in level ℓ.

22.4 A reduction of polynomial calculus PC to NS

The Nullstellensatz is a proof system for proving that the ideal generated by some polynomials is trivial. One can take the closure properties of ideals as inference rules and define a different proof system, the so-called polynomial calculus PC. We shall now recall its definition (just for the ring S).

Definition 22.4.1 (Clegg *et al.* [24])
Let $f_1,\dots,f_k,g$ be polynomials over S. A **polynomial calculus proof** (PC-proof) of g from $f_1,\dots,f_k$ is a sequence of polynomials $h_1,\dots,h_t$ from S such that $h_t = g$, and for each $u \le t$:

1. either h_u is one of $f_1,\dots,f_k$,
2. or $h_u = h_v + h_w$ for some $v, w < u$,
3. or $h_u = h_v \cdot x_{ij}$, for some $v < u$ and x_{ij} a variable.

The degree of the PC-proof is $\max_{u \le t} \deg(h_u)$.

We write $f_1,\dots,f_k \vdash_d g$ to indicate that there is a PC-proof of g from $f_1,\dots,f_k$ of degree at most d.

Theorem 22.2.4 provided a strong degree lower bound for NS refutations of $(\neg\mathrm{PHP}_n)$. Razborov [94] proved a similar lower bound for PC. Note that a degree d NS-refutation can be easily turned into a degree d PC-refutation. The opposite is not true in general but Razborov's construction actually shows that, as far as proofs from $(\neg\mathrm{PHP}_n)$ are concerned, PC is no stronger than NS (see Razborov [94, Remark 3.12]).

Theorem 22.4.2 (Razborov [94]) *There is no degree $n/2$ PC-proof of* 1 *from* $(\neg\mathrm{PHP}_n)$ *in S.*

For any $f \in S$, if f has a degree $d \le n/2$ PC-proof from $(\neg\mathrm{PHP}_n)$ *then it also has a degree d NS-proof from* $(\neg\mathrm{PHP}_n)$.

Having such a strong degree lower bound it would seem that we can substitute in the role of samples for the desirable but non-existent maximal ideals maximal subsets of S, containing $(\neg\mathrm{PHP}_n)$, closed under degree $n/2$ PC-derivations but not containing 1. Such sets do exist and their intersections with S^{low} are, in fact, ideals in S^{low}. However, again no such set is definable in $\mathcal{M}$ and, in particular, the set of such samples would not be in $\mathcal{M}$, violating a key requirement of forcing with random variables.

We now outline Razborov's construction in some detail as it provides an explicit description of a certain vector space basis that may turn out to be eventually useful. We follow Razborov [94] and, in parts, a simplified treatment by Impagliazzo, Pudlák and Sgall [46]. Our notation attempts to adhere to these two papers.

We shall define first several ambient vector spaces (everything is over $\mathbf{F}_2$) and describe a certain vector space basis. The definition of the pigeon dance and the basis Δ_t is due to Razborov [94], the basis C_t was explicitly defined by Impagliazzo *et al.* [46].

Definition 22.4.3

1. $\hat{S}$ is the ring S factored by the ideal generated by all polynomials $Q_{i_1,i_2;j}$ and $Q_{i;j_1,j_2}$ from $(\neg\mathrm{PHP}_n)$.
   For $f \in S, \hat{f} \in \hat{S}$ denotes the corresponding element of the quotient.
2. Map is the set of partial injective maps a from $[n+1]$ into $[n]$. Each $a \in \mathsf{Map}$ defines a monomial $x_a := \Pi_{i \in \mathrm{dom}(a)} x_{ia(i)}$. In particular, $\emptyset \in \mathsf{Map}$ and $x_\emptyset = 1$. For $t \geq 0$, put $\mathsf{Map}_t := \{a \in \mathsf{Map} \mid |a| \leq t\}$.
3. For $t \geq 0$, $\hat{S}_t$ is the set of elements of $\hat{S}$ of degree at most t.
   $T_t := \{x_a \mid a \in \mathsf{Map}_t\}$ is the vector space basis of $\hat{S}_t$.
4. V_t is the set of polynomials $\hat{f} \in \hat{S}$ such that $f \in S$ has a degree $\leq t$ PC-proof (equivalently, by Theorem 22.4.2, a degree $\leq t$ NS-proof) from $(\neg\mathrm{PHP}_n)$.
5. Let $a \in \mathsf{Map}$. A **pigeon dance** of a is the following non-deterministic process: take the smallest pigeon $i_1 \in \mathrm{dom}(a)$ and move it to any currently unoccupied hole bigger than $a(i_1)$. Then take the second smallest pigeon $i_2 \in \mathrm{dom}(a)$ and move it to any currently unoccupied hole bigger than $a(i_2)$, etc. We say that the pigeon dance is defined on a if this process can be completed for all pigeons in $\mathrm{dom}(a)$.
6. Map^* is the set of all partial maps $a : [n+1] \to [n] \cup \{0\}$ that is injective on $\mathrm{dom}(a) \setminus a^{(-1)}(0)$. $\mathsf{Map}^*_t := \{a \in \mathsf{Map}^* \mid |a| \leq t\}$.
   For $a \in \mathsf{Map}^*$ denote by a^- the map a restricted to domain $\mathrm{dom}(a) \setminus a^{(-1)}(0)$. In particular, $a^- \in \mathsf{Map}$.
   For $a \in \mathsf{Map}^*$ denote by x_a the element of $\hat{S}$ represented by $x_{a^-} \cdot Q_{i_1} \cdot \dots \cdot Q_{i_k}$, where $\{i_1, \dots, i_k\} = a^{(-1)}(0)$.
7. $B_t \subseteq \hat{S}$ is the set of all elements x_a for $a \in \mathsf{Map}^*_t$ such that a pigeon dance is defined on a^-. In particular, $1 \in B_0$.
8. Put $C_t := B_t \setminus T_t$ and $\Delta_t := B_t \cap T_t$. In particular, Δ_t is the set of all elements x_a for $a \in \mathsf{Map}_t$ such that a pigeon dance is defined on a.

Theorem 22.4.4 (Razborov [94], Impagliazzo *et al.* [46])

(1) *For $0 \leq t \leq n/2$, as vector spaces*

$$\hat{S}_t = V_t \oplus \mathbf{F}_2 \Delta_t$$

and

$$V_t = \mathbf{F}_2 C_t.$$

In particular, $1 \notin V_t$ *and* Δ_t *is a basis of the vector space* $\hat{S}_t / V_t$.

(2) *Let* $x_c \in T_t$ *and assume that its (unique) expression in the basis* $C_t \cup \Delta_t$ *is*

$$\sum_{x_a \in X} \alpha_a x_a + \sum_{x_b \in Y} \alpha_b x_b$$

for some $X \subseteq C_t$ *and* $Y \subseteq \Delta_t$, *and* α*s non-zero coefficients from* $\mathbf{F}_2$. *Then, for all* $x_a \in X$ *and* $x_b \in Y$, $\mathrm{dom}(a) \subseteq \mathrm{dom}(c)$ *and* $\mathrm{dom}(b) \subseteq \mathrm{dom}(c)$.

(3) *Let* $f_0, f_1 \in \hat{S}$, $\deg(f_0) + \deg(f_1) \leq t$ *and* $t \leq n/4$. *If* f_0 *has a degree* $\leq t$ *Ns-proof from* $(\neg PHP_n)$ *then* $f_0 \cdot f_1 \in V_t$.

The first statement of part (1) of the theorem is from Razborov [94] (Claims 3.4 and 3.11 there, showing that Δ_t is a basis of $\hat{S}_t/V_t$). An alternative proof of that and the second statement in part (1) are in Impagliazzo *et al.* [46] (Proposition 3.8 and the proof of Theorem 3.9 there).

Part (2) just states the property of the reduction process from Razborov [94] (the proof of Claim 3.4 there, repeated in the proof of Proposition 3.8 in Impagliazzo *et al.* [46]) that finds the expression of x_c in terms of basis $C_t \cup \Delta_t$. We now give a brief sketch of this reduction. This will prove part (2) of the theorem and, in fact, it also shows that $C_t \cup \Delta_t$ spans $\hat{S}_t$ and that Δ_t spans $\hat{S}_t/V_t$ (the linear independence needs an additional argument). The reduction process also yields part (3).

Define a partial ordering $x_c \preceq x_d$ for terms $x_c = x_{i_1 j_1} \dots x_{i_k j_k}$ and $x_d = x_{u_1 v_1} \dots x_{u_\ell v_\ell}$ from T_t by:

- either $k < \ell$, or
- $k = \ell$ and for the largest w such that $j_w \neq v_w$ it holds that $j_w < v_w$.

The statement of part (2) of the theorem is proved by induction on $\preceq$. Let $x_c = x_{i_1 j_1} \dots x_{i_k j_k}$ with $i_1 < \dots < i_k$. Assume $x_c \notin \Delta_t$ (otherwise there is nothing to prove) and that the statement holds for all terms $\prec$-smaller than x_c. Rewrite x_c using axiom Q_{i_1} as

$$x_{i_2 j_2} \dots x_{i_k j_k} + \sum_{j_1^* < j_1} x_{i_1 j_1^*} \dots x_{i_k j_k} + \sum_{j_1^* > j_1} x_{i_1 j_1^*} \dots x_{i_k j_k}$$

(where we automatically delete terms with $j_1^* \in \{j_2,\dots,j_k\}$). The first term as well as the terms in the first summation are all $\prec$-smaller than x_c and their domain is included in $\mathrm{dom}(c)$. The statement for them follows by the induction hypothesis.

The terms in the last summation can be thought of as all possible moves of pigeon i_1 in the attempted dance of c. Now rewrite each of these terms analogously with respect to pigeon i_2, etc. But as $x_c \notin \Delta_t$ the pigeon dance cannot be completed and this rewriting will eventually produce only terms $\prec$-smaller than x_c.

22.5 The necessity of partially defined random variables

In this section we leave the specifics related to Problem 22.0.1 and pause to reflect on the differences in the construction of Skolem functions in $K(F_{\mathrm{PHP}}, G_{\mathrm{PHP}})$ from those in structures $K(F_{\mathrm{tree}}, G_{\mathrm{tree}})$ and $K(F_{\mathrm{alg}}, G_{\mathrm{alg}})$. One may expect to tackle similar issues in the eventual construction of the desired model of $Q_2V_1^0$ in which PHP fails.

Some differences in the constructions of Skolem functions in these structures were a priori necessary as Skolem functions in $K(F_{\mathrm{tree}}, G_{\mathrm{tree}})$ and $K(F_{\mathrm{alg}}, G_{\mathrm{alg}})$ are in a sense 'too good': they imply not only bounded quantifier elimination but also the preservation of true $s\Pi_1^{1,b}$-sentences. This is a property that no model that is hoped to have proof complexity applications can have, as we have discussed in Part V.

The qualification 'too good' means that structures $K(F_{\mathrm{tree}}, G_{\mathrm{tree}})$ and $K(F_{\mathrm{alg}}, G_{\mathrm{alg}})$ have Skolem functions providing bounded quantifier elimination in the following very strong sense. Let $B(x,X)$ be a $\Sigma_0^{1,b}$-formula. Then for any $m \in \mathcal{M}_n$ and any Γ there is an open formula $C(x)$ such that it not only holds that

$$(1) \qquad [\![\forall x < m\, B(x,\Gamma) \equiv C(x)]\!] \;=\; 1_{\mathcal{B}}$$

but even for any α

$$(2) \qquad \langle\!\langle \alpha < m \to B(\alpha,\Gamma)\rangle\!\rangle \;\Delta\; \langle\!\langle \alpha < m \to C(\alpha)\rangle\!\rangle \;\in\; \mathcal{I}.$$

In other words, the symmetric difference has an infinitesimal counting measure. Property (2) is what implies the preservation of true $s\Pi_1^{1,b}$-sentences and what therefore had to fail in $K(F_{\mathrm{PHP}}, G_{\mathrm{PHP}})$, and must do so in any other model with proof complexity ambitions. Recall briefly the argument from Section 12.2: let $A(x)$ be of the form $\forall X < x B(x,X)$, formula $B(x,X)$ as above. Assuming that

$\forall x A(x)$ is true in $\mathbf{N}$ (ignoring $X < x$) then for all m and any Γ it holds that

$$\langle\langle \alpha < m \rightarrow B(\alpha, \Gamma)\rangle\rangle \; = \; \Omega$$

and hence, by (2),

$$\Omega \setminus \langle\langle \alpha < m \rightarrow C(\alpha)\rangle\rangle \; \in \mathcal{I},$$

which implies, as $C(x)$ is open, that $[\![\alpha < m \rightarrow C(\alpha]\!] = 1_{\mathcal{B}}$ and hence, by (1), $[\![\alpha < m \rightarrow B(\alpha, \Gamma)]\!] = 1_{\mathcal{B}}$ too. Note for the record that a completely analogous argument shows that the existence of Skolem functions with properties (1) and (2) implies that for any $\Sigma_0^{1,b}$-sentence D with parameters it holds that

$$[\![D]\!] \; = \; \langle\langle D\rangle\rangle \; / \; \mathcal{I}.$$

Namely, by (1), $[\![D]\!] \; = [\![D']\!]$ for some quantifier-free sentence D', for which therefore $[\![D']\!] = \langle\langle D'\rangle\rangle / \mathcal{I}$ holds, and (2) thus implies the observation.

Let us try next to understand in general terms where there was any room in the construction of $K(F_{\text{PHP}}, G_{\text{PHP}})$ for the failure of property (2).

The switching lemma for PHP-trees (Theorem 21.1.1) is more subtle and has a harder proof than the original switching lemma in the plain Boolean case (Theorem 10.1.2). This is, however, not the only difference between the analysis of structures $K(F_{\text{tree}}, G_{\text{tree}})$ and $K(F_{\text{PHP}}, G_{\text{PHP}})$. The additional difficulty stemmed from the fact that random variables involved in the definition of $K(F_{\text{PHP}}, G_{\text{PHP}})$ were only partially defined. Hence there were suddenly two sources of imprecision in the construction of Skolem functions: one had to do with the fact that the switching lemma works only with some (albeit exponentially small) error; the other one had to do with the regions of undefinability of random variables in question. The former property is shared by the Razborov–Smolensky approximation method; it too introduces an exponentially small error. As it turned out, it is exactly the latter property, the seemingly innocent fact that we allow partially defined random variables, that caused the failure of property (2) above.

The idea of the constructions of Skolem functions in Theorems 11.1.1 and 15.1.1 is the same although its technical implementations are different in both cases (the use of the switching lemma and the Razborov–Smolensky approximation, respectively). Also the analysis of why the construction actually works is the same in both cases (based on bounding the total size of the error sets). The construction of Skolem functions for $K(F_{\text{PHP}}, G_{\text{PHP}})$ was again formally the same as before (utilizing the PHP switching lemma, i.e. Theorem 21.1.1) but what was different was the analysis of why it actually worked.

Let us discuss this point in some detail and again in general terms, as it may shed light on the desired construction. First an example: as pointed out earlier, property (2) of Skolem functions implies, in particular, that for an open formula $C(x)$ the truth value $[\![\exists x < m\ C(x)]\!]$ is equal to $\langle\langle \exists x < m\ C(x)\rangle\rangle/\mathcal{I}$. Even this corollary of (2) is no longer true in $K(F_{\mathrm{PHP}}, G_{\mathrm{PHP}})$. Take the formula $\exists x \in [n+1]\ \Delta(x) \notin [n]$. Clearly $\langle\langle \exists x \in [n+1]\ \Delta(x) \notin [n]\rangle\rangle = \Omega_{\mathrm{PHP}}$ but by the analysis in Lemma 21.3.3 $[\![\exists x \in [n+1]\ \Delta(x) \notin [n]]\!] = 0_{\mathcal{B}}$. The set $\langle\langle \exists x \in [n+1]\ \Delta(x) \notin [n]\rangle\rangle$ is equal to $\bigcup_{i \in [n+1]} \langle\langle \Delta(i) \notin [n]\rangle\rangle$. By definition, $\Delta(i)$ is the depth 1 PHP-tree querying $i \to ?$ and with the leafs labeled by $j \in [n]$. Hence $\langle\langle \Delta(i) \in [n]\rangle\rangle$ is represented by the same tree with all labels changes to 1 (yes). Then also $\langle\langle \Delta(i) \notin [n]\rangle\rangle$ is represented by the same tree but now with all labels 0 (no). Hence the eventual Skolem function looks at a disjunction of no's and cannot find any witness for the formula.

Let us now briefly repeat the analysis of the earlier constructions and point out where we needed to be more subtle in the case of a Skolem function for $K(F_{\mathrm{PHP}}, G_{\mathrm{PHP}})$ than before. In the constructions of a Skolem function for an open formula $B(x,y)$ on m we seek first to find β_i such that

$$[\![\exists y < mB(i,y)]\!] = [\![\beta_i < m \wedge B(i,\beta_i)]\!]$$

and then take for the Skolem function the function determined by the tuple $(\beta_i)_{i<m}$. For this to work we need to pick β_i in such a way that, given a sample, no α can find with more than an infinitesimal probability an i such that the sample is in the error set

$$E_i := \langle\langle \exists y < mB(i,y)\rangle\rangle \setminus \langle\langle \beta_i < m \wedge B(i,\beta_i)\rangle\rangle.$$

In the proofs of Theorems 11.1.1 and 15.1.1 this is achieved by finding β_i such that each E_i has not only an infinitesimal counting measure but even an exponentially small measure. This then guarantees that their union has an infinitesimal counting measure and that is what is aimed at.

In the case of $K(F_{\mathrm{PHP}}, G_{\mathrm{PHP}})$ this analysis would have run into a problem. Each β_i is undefined on a set whose counting measure is infinitesimal but not exponentially small. The region of undefinability of each β_i may have the measure $\Omega(n^{-\Omega(1)})$ so already $n^{O(1)}$ of these sets can sum up to a set covering the whole sample space (and m can be even much bigger than that).

Fortunately a more careful analysis (that would work in earlier cases too but was not needed) showed that this can be circumvented. Recall the idea here using a simple example, an existential sentence. Let $C(y)$ be an open formula (possibly with parameters) and $m \in \mathcal{M}_n$. Assume we want to find β (a

0/1-valued PHP-tree) that signals the validity of $\exists y < m\, C(y)$; that would be the first step, analogously with earlier constructions of Skolem functions, in a construction of a Skolem constant γ such that:

$$[\![\exists y < m\, C(y)]\!] = [\![\gamma < m \wedge C(\gamma)]\!].$$

Consider 0/1-valued PHP-trees S_j representing $\langle\!\langle C(j)\rangle\!\rangle$, for $j < m$. Let D_j be the disjunction of map-terms corresponding to paths in S_j labeled by 1, and let $D := \bigvee_{j<m} D_j$. Assume PHP-tree R represents D. We would like to argue that β defined by R works for our purposes.

The first argument that comes to mind, rather as in earlier constructions of Skolem functions, is to argue that R represents $\bigcup_{j<m}\langle\!\langle C(j)\rangle\!\rangle$. We ignore the sample (an infinitesimal fraction of all samples) where β is undefined. If $\beta(T)$ is defined and $\beta(T) = 1$, then $D(T) = 1$ and hence also $D_j(T) = 1$ for some $j < m$. By Lemma 20.1.5 also $S_j(T) = 1$ and hence $T \in \langle\!\langle C(j)\rangle\!\rangle$. Note that this in itself does not necessarily imply anything like

$$[\![\beta = 1]\!] \le [\![\exists y < m\, C(y)]\!].$$

For this one needs to complete the construction of γ in the binary tree fashion employed in the earlier constructions. We will not discuss this case here in detail (see Section 21.4) as the second case is more interesting for our general discussion.

In the second case $\beta(T)$ is defined but equal to 0. Then $D(T) = 0$ and all $D_j(T) = 0$. However, $D_j(T) = 0$ does not imply that also $S_j(T) = 0$ (which would yield the desired $T \notin \langle\!\langle C(j)\rangle\!\rangle$). It only implies that either $S_j(T) = 0$ or T is in the region of undefinability of S_j. And we have no a priori control over how large the union of these individual regions is; it can be the whole sample space.

The argument that did work was slightly more subtle. Instead of arguing that β represents $\bigcup_{j<m}\langle\!\langle C(j)\rangle\!\rangle$ we argued that, for any α, β represents a predicate containing $\langle\!\langle \alpha < m \wedge C(\alpha)\rangle\!\rangle$.

Ignoring the $\alpha < m$ part, $\langle\!\langle C(\alpha)\rangle\!\rangle$ is represented by a PHP-tree $\Gamma(\alpha)$, where Γ is the map defined by the m-tuple $(S_j)_{j<m}$. So it suffices to verify that

$$\Gamma(\alpha)(T) = 1 \rightarrow \beta(T) = 1$$

if they are both defined. If $\Gamma(\alpha)(T) = 1$ then $\alpha(T)$ is defined and equal to some $j < m$, and $S_j(T)$ is defined and equal to 1. So $D_j(T) = 1$ and $D(T) = 1$, and $\beta(T) = 1$ too. The point is, however, that the set of T for which $\Gamma(\alpha)$ or β are undefined is of an infinitesimal measure.

In other words, we relied in Section 21.4 on a general fact that if we have PHP-trees α and S_j for $j < m$, all of depth at most d, and we attach trees S_j arbitrarily to the leafs of α then the region of undefinability of the resulting tree is not the sum of the regions of undefinability of the individual trees. It is much smaller; its counting measure is only $O(2d/n)$. This is due to the key Lemma 20.1.3.

In the light of the constructions of the earlier structures and the above discussion we may search for a structure constructed as follows. The sample space Ω is formed by a suitable set of some (not all) partial one-to-one maps from $[n+1]$ to $[n]$. Note that all $\omega \in \Omega$ satisfy all equations defining $\hat{S}$.

The random variables underlying the structure would be defined by some algebraic trees. It seems that in the algebraic context the notation introduced after Lemma 14.1.4 is more natural. Consider an expression

$$\alpha := \sum_{i \in u} p_i \cdot i^*$$

where $u \in \mathcal{M}_n$, all $p_i \in \hat{S}^{\text{low}}$ satisfying in the ring

$$p_i \cdot p_j = 0, \quad \text{for } i \neq j < u.$$

This defines a map from Ω to $\mathcal{M}_n$: $\alpha(\omega)$ equal to the unique $i < u$ such that $p_i(\omega) = 1$ if it exists, and is undefined if all $p_i(\omega) = 0$. Hence α is possibly only partially defined. Denote by $\alpha(\omega) \uparrow$ the fact that α is undefined at ω.

The crucial condition to arrange, analogous to key Lemma 20.1.3, is that

$$\text{Prob}_{\omega \in \Omega}[\alpha(\omega) \uparrow] \textit{ is infinitesimal.}$$

One needs to arrange a few simple closure properties of family F in order to show that open comprehension and open induction are valid in the structure, and that the construction of Skolem functions works as in $K(F_{\text{alg}}, G_{\text{alg}})$.

There is a natural choice for a map Δ, a second-order object in the structure, that would violate PHP in the structure. Take

$$\Delta := (\beta_1, \ldots, \beta_{n+1})$$

where

$$\beta_i := \sum_{j \in [n]} x_{ij} \cdot j^*.$$

This is quite analogous to $K(F_{\text{PHP}}, G_{\text{PHP}})$. The requirement is then that F is closed under Δ.

All these conditions seem to require fine balancing to apply at the same time.

PART VII

Polynomial-time and higher worlds

23
Relevant theories

We shall turn in this part to theories adequate for polynomial time or stronger computational models. These include Cook's theory PV, Buss's theory S_2 and, in particular, its subtheories S^1_2, T^1_2, and theory $BT = (S^1_2 + dW\text{PHP}(\Delta_1))$. We shall use mostly the first-order formulation of these theories (with the exception of theories U^1_1 and V^1_1). We give definitions of the theories but otherwise do not recall much of a background that is not directly used later. The reader may find a much more detailed exposition in Krajíček [56].

Some of the first-order theories are presented in the second-order formalism in Cook and Nguyen [30].

23.1 Theories PV and $\text{Th}_\forall(L_{\text{PV}})$

Cook's theory PV (standing for Polynomially Verifiable) was a first theory that we today include in the collection of bounded arithmetic theories. Its original definition in Cook [28] presented PV as an equational theory but we shall follow later presentations and define the theory as first order (this is often called PV_1 but we will abuse the notation slightly and stick just with PV).

Roughly speaking the theory uses a language with a name for every polynomial-time algorithm, and its axioms codify how these algorithms are built one from another. The names for polynomial-time algorithms are introduced following Cobham's [26] characterization of polynomial-time functions on **N**. This characterization says that the class of polynomial-time functions is the minimal class of functions $\mathcal{C}$ such that

1. constant 0 and functions $s_0(x)$, $s_1(x)$ and $x\#y$ are in $\mathcal{C}$, where

$$s_0(x) := 2x, \quad s_1(x) := 2x+1 \text{ and } x\#y := 2^{|x|\cdot|y|}$$

   (with $|x|$ being the bit length of number x),

2. C is closed under permutation of arguments of functions and under compositions, and
3. C is closed under limited recursion on notation.

The limited recursion on notation defines a new function $f(\overline{x}, y)$ from functions g, h_0, h_1, ℓ if the following hold:

(a) $f(\overline{x}, 0) = g(\overline{x})$.
(b) $f(\overline{x}, s_i(y)) = h_i(\overline{x}, y, f(\overline{x}, y))$, for $i = 0, 1$.
(c) $f(\overline{x}, y) \leq \ell(\overline{x}, y)$.

The language L_{PV} of PV has names for the functions from condition 1, as well as names for functions introduced by conditions 2 and 3, and relation $\leq$ (and equality which is included automatically). Axioms of PV are universal formulas saying how a function was introduced, and the scheme of induction for open formulas. For example, if f was introduced by the limited recursion on notation as above, PV has as axioms the four formulas in (a), (b) and (c).[1]

We remark that any open induction axiom can be also formulated as a universal formula using a function (with a name in L_{PV}) simulating the binary search (see Krajíček [56] for details). Thus PV is a universal theory.

Note that the open induction suffices to prove that if functions f and f' were introduced by the limited recursion on notation from two sets of functions g, h_0, h_1, ℓ and g', h'_0, h'_1, ℓ' respectively, such that $g = g'$, $h_i = h'_i$ and $\ell = \ell'$, then also $f = f'$.

Theory $\mathrm{Th}_{\forall}(L_{\mathrm{PV}})$ is axiomatized by all true universal sentences in L_{PV}. Of course, we have no explicit description of what these sentences are but the theory is useful in connection with witnessing theorems.

23.2 Theories S^1_2, T^1_2 and BT

The language of Buss's theory T_2 extends language L_{PA} (see Section 5.1) by function symbols

$$|x|, \quad \lfloor \frac{x}{2} \rfloor \quad \text{and} \quad x\#y.$$

The theory is axiomatized by a set of axioms (called BASIC) defining 'basic' properties of the symbols in the language, and by the scheme of induction IND for bounded formulas in the language (as in Chapter 6).

BASIC naturally extends set PA^- from Chapter 6 by adding axioms concerning the new function symbols. We shall not spell them out (see Buss [11] or Krajíček [56]). In fact, one often takes for a convenience L_{PV} as the language.

[1] One has to be a bit more subtle to get a polynomial-time set of axioms; see Cook [28] or Krajíček [56].

Then all axioms of PV are also included in BASIC and the theory is denoted $T_2(\mathrm{PV})$.

Buss [11] considered also two other forms of induction: the LIND scheme (Length IND),

$$A(0,\overline{y}) \wedge \forall x(A(x,\overline{y}) \rightarrow A(x+1,\overline{y})) \rightarrow \forall x A(|x|,\overline{y}),$$

and the PIND scheme (Polynomial IND),

$$A(0,\overline{y}) \wedge \forall x \left(A\left(\lfloor \frac{x}{2} \rfloor,\overline{y}\right) \rightarrow A(x,\overline{y})\right) \rightarrow \forall x A(x,\overline{y}).$$

He proved that when the schemes are accepted for all bounded formulas they both define (over BASIC) theories equivalent to T_2. S_2 is the theory based on PIND for bounded formulas. Hence $T_2 = S_2$.

However, fragments of these theories, obtained by restricting the schemes IND and PIND to bounded formulas of certain complexity, are not known to be equivalent (and are not expected to be).

The relevant hierarchies Σ_i^b and Π_i^b, $i = 0, 1, \ldots$, of bounded formulas are defined completely analogously with the arithmetical hierarchy Σ_i^0, with the roles of bounded and unbounded quantifiers in the definition of the latter hierarchy taken by sharply bounded quantifiers $\exists y < |t|$ and $\forall y < |t|$ (the variables bounded above by the length of a term) and by bounded quantifiers respectively. In particular, Σ_0^b-formulas contain only sharply bounded quantifiers (in a similar way as Σ_0^0 contains only bounded ones) and Σ_1^b include all formulas resulting from Σ_0^b-formulas by prefixing a bounded existential quantifier while Π_1^b includes all formulas resulting from Σ_0^b-formulas by prefixing a bounded universal quantifier.

Theories T_2^i and S_2^i are then subtheories of T_2 and S_2 obtained by restricting the schemes IND and PIND, respectively, to Σ_i^b-formulas only.

Denote by $S_2^i(\mathrm{PV})$ the theory with L_{PV} as its language. A key fact proved by Buss (see Buss[11] or Krajíček [56]) is that $S_2^1(\mathrm{PV})$ is conservative over PV for all Σ_1^b-formulas. More results about the theories will be recalled when needed.

23.3 Theories U_1^1 and V_1^1

Theories U_1^1 and V_1^1 have the same language as V_1^0. Theory V_1^1 extends V_1^0 by accepting the scheme of induction IND for all $\Sigma_1^{1,b}$-formulas (recall these formulas from Section 5.1).

Theory U^1_1 extends V^0_1 by accepting instead the scheme of polynomial induction PIND for all $\Sigma^{1,b}_1$-formulas. It is known that U^1_1 is contained in V^1_1.

I shall say more about the theories when we need it but now let us recall only one simple model-theoretic correspondence between V^1_1 and S^1_2. Having a model $\mathbf{A} = (A, \mathcal{A})$ of V^1_1, with A the numbers and $\mathcal{A}$ the bounded sets of the model, we may construct a model B of S^1_2 as follows: the elements (numbers) of B are represented by $\mathcal{A}$, thinking of a set as a string of bits defining a number. One can then define in $\mathbf{A}$ the interpretation of the language of S^1_2 on B and the structure will be a model of S^1_2.

On the other hand, given a model B of S^1_2 take for A the sharply bounded numbers of B, i.e. the lengths of elements of B, and for $\mathcal{A}$ the (necessarily bounded) subsets of B coded in B. The resulting structure will satisfy V^1_1. See Krajíček [56] for details.

24

Witnessing and conditional independence results

In this chapter we give new proofs of some of the known witnessing theorems and conditional independence results in bounded arithmetic. We take, for the sake of example, the following three important results, but it will be clear how to reprove other results of this sort in the same way. The three results are:

(1) PV $\neq S^1_2$, assuming that NP $\not\subseteq$ P/poly (see Krajíček, Pudlák and Takeuti [78]).
(2) $S^1_2 \neq T^1_2$, assuming that $\mathrm{P}^{\mathrm{NP}} \neq \mathrm{P}^{\mathrm{NP}}[O(\log n)]$ (see Krajíček [54]).
(3) S^1_2 does not prove the weak pigeonhole principle for polynomial-time functions, assuming that RSA is secure or assuming that strong collision-free hash families exists (see Krajíček [61] and Krajíček and Pudlák [76]).

The qualification *new* means that statements of Section 2.2 and Chapter 3 replace the use of proof theory in the original arguments. A price worth paying for this is that we need to modify the complexity assumptions from the worst-case hardness to the average-case hardness.

We formulate the respective witnessing theorems only as parts of proofs of the independence results. One reason is that they are immediate consequences of the general facts from Chapter 3. Another reason for not trying to formulate them universally is that in the coming constructions sample spaces vary from case to case and are always tailored to the particular independence results.

It will be clear that analogous results for higher theories S^i_2 and T^i_2 with $i > 1$ can be proved identically by simply allowing the random variables in the particular L-closed families to be sampled by algorithms of the type used in the base case $i = 1$ but with an access to an oracle from an appropriate level of the polynomial-time hierarchy.

We will treat the three statements in reverse order, from simpler constructions to harder ones. Recall that L_{PV} is the language of theory PV, having a name for every polynomial-time algorithm (see Section 23.1).

The definitions of the models follow the original first-order version of Section 1.3. In particular, the cut $\mathcal{M}_n$ makes no (explicit) appearance.

24.1 Independence for S^1_2

Let WPHP(C,a) be the existential $L_{\rm PV}$-formula formalizing the **weak pigeonhole principle**:

- *Either C is not a circuit computing a map from $\{0,1\}^{|a|+1}$ into $\{0,1\}^{|a|}$ or there are $u \neq v \in \{0,1\}^{|a|+1}$ such that $C(u) = C(v)$.*

For a number k let RSA$_k$ denote the set of all pairs (g,N) such that

- N is a product of two primes (denoted p and q) of length k (i.e. $|N| = 2k$)
- $1 < g < N$ and g is coprime to $(p-1)(q-1)$.

By an RSA function based on such a pair (g,N) we mean a function

$$x < N \to g^x \bmod N.$$

The following theorem alludes to statement (3).

Theorem 24.1.1 *Assume that there is a standard $\epsilon > 0$ such that no polynomial-time algorithm can find the secret key for more than $(1-\epsilon)$ fraction of RSA functions based on a pair from* RSA$_k$*, for all k large enough.*

Then S^1_2 (even augmented by Th$_{\forall_1}(L_{\rm PV})$*) does not prove the sentence*

$$\forall C, a \ {\rm WPHP}(C,a).$$

Proof: Consider the set of pairs RSA$_{n/2}$, n our non-standard parameter, as the sample space. Let $F_{\rm PV}$ be the family of polynomial-time computable functions with domain RSA$_{n/2}$.

Let $\alpha \in F_{\rm PV}$ be a function that from sample $(g,N) \in {\rm RSA}_{n/2}$ computes a circuit $C_{g,N}$ computing the map $x \in \{0,1\}^{n+1} \longrightarrow g^x \bmod N \in \{0,1\}^n$. Note that the statement 'any $C_{g,N}$ is a circuit with $n+1$ inputs and n outputs' is $K(F_{\rm PV})$-valid as it is a true universal statement (Lemma 1.4.2). For the same reason PV is valid in the structure.

Witnessing claim: *Assume that $\forall C, a$* WPHP(C,a) *is $K(F_{\rm PV})$-valid. Then for any standard $\epsilon > 0$ there is a $\beta \in F_{\rm PV}$ such that:*

$$\mu([\![\beta \textit{ is a pair } u \neq v \in \{0,1\}^{n+1} \textit{ s.t. } \alpha(u) = \alpha(v)]\!]) \geq 1 - \epsilon.$$

In particular,

$$\text{Prob}_{(g,N)}[\ \beta(g,N) \textit{ is a pair } u \neq v \in \{0,1\}^{n+1} \textit{ s.t. } g^u \equiv g^v \bmod N\ ] \geq 1-\epsilon.$$

If the sentence was valid then after instantiating the universal quantifiers by α for C and by 2^n for a Lemma 3.3.2 guarantees the existence of the required β, for any standard $\epsilon > 0$. The inequality for the probability follows then by Lemma 2.2.1.

However, a pair of $u \neq v$ provided by the Claim gives $w := u - v \neq 0$ such that $g^w \equiv 1 \bmod N$, and it is well known that having such w is enough to break the particular RSA. Hence, using β, we could do it for at least a $(1-\epsilon)$ fraction of all parameters from RSA_n, for any standard $\epsilon > 0$. That contradicts the hypothesis of the theorem.

The theorem follows by Lemma 1.4.1 noting that S^1_2 is $\forall\exists$-conservative over PV (by Buss [11]) which is $K(F_{\text{PV}})$-valid as it is a true universal theory (Lemma 1.4.2). The same argument shows that we can allow $\text{Th}_{\forall_1}(L_{\text{PV}})$ as extra axioms. □

The alternative hypothesis in statement (3) about hash families can be treated analogously (see Krajíček [61]).

We leave it to the reader to verify that one can similarly prove, assuming the existence of strong one-way permutations, that S^1_2 does not prove the following principle for polynomial-time functions:

- *Any injective map* $f : \{0,1\}^n \longrightarrow \{0,1\}^n$ *must be onto.*

24.2 S^1_2 versus T^1_2

Next we turn to statement (2). Let Ξ be a family of counter-example functions (see Section 2.2), one for each formula $\forall z \leq t(x) A(x,z)$, A an open L_{PV}-formula and $t(x)$ an L_{PV}-term.

We shall denote by $\text{FP}^{\Xi}[q(n)]$ the class of functions computable by a polynomial-time oracle machine that is allowed at most $q(n)$ queries to a function from Ξ.

Note that a query to an NP-oracle $A(u) = \exists v \leq t(u) B(u,v)$ can be answered if we have access to a counter-example function for $\neg A(u)$. Hence the class $\text{FP}^{\Xi}[q(n)]$ contains the bounded query classes $\text{FP}^{\text{NP}}[q(n)]$ of Buss and Hay [17] and Wagner [107] and, in fact, also the bounded query classes with witness-oracles $\text{FP}^{\text{NP}}[\text{wit}, q(n)]$ of Buss, Krajíček and Takeuti [20] and Krajíček [54].

Theorem 24.2.1 *Assume that there is a* P^{NP} *function such that no bounded query* $P^{NP}[O(\log n)]$ *algorithm can compute the function on more than* $(1-\epsilon)$ *fraction of all inputs, some* $\epsilon > 0$.

Then $S^1_2 \neq T^1_2$.

Proof: Let F be the family of functions from $\bigcup_c FP^{\Xi}[c \cdot \log(n)]$ with domain $\Omega := \{0,1\}^n$, for some non-standard n, where the constant c in the union ranges over standard $c \geq 1$.

Claim 1: S^1_2 *is* $K(F)$*-valid.*

F is an L_{PV}-closed family, so PV is $K(F)$-valid. The claim follows (via Lemma 1.4.1) from the fact that S^1_2 is axiomatized over PV by the sharply bounded function minimization scheme $\#FM\,(C,a)$:

- *If* C *is a circuit computing on* $\{0,1\}^{|a|}$ *a function whose values are numbers* $< |a|$ *then the minimal value is assumed on some* $u \in \{0,1\}^{|a|}$;

see Chiari and Krajíček[23] and Hanika [39].

Clearly the minimal value $t < |a|$ can be found by binary search with $\log|a| = O(\log n)$ (as a priori $|a| = n^{O(1)}$) queries to an NP-oracle, and then a suitable u can be found by one call to a counter-example function for formula $\forall z(|z| \leq |a|)(f(z) \geq t+1)$. Hence the principle $\forall C, a\ \#FM\,(C,a)$ is witnessed by a function in F and so is, by Lemma 1.4.2, $K(F)$-valid (substituting a witnessing function for the existential quantifier in an $\forall\exists\forall$-sentence transforms it into a true universal sentence).

Witnessing claim 2: *Assume* T^1_2 *is* $K(F)$*-valid. Then for any* P^{NP} *function* $f(x)$ *there is* $\beta \in F$ *such that:*

$$\operatorname*{Prob}_{\omega}[\beta(\omega) = f(\omega)] \geq 1-\epsilon,$$

where ω *ranges over* Ω.

By Buss [11] the graph $f(x) = y$ of any P^{NP} function can be defined by an $\exists\forall$ L_{PV}-formula $\exists z \forall t A(x,y,z,t)$ such that T^1_2 proves that it is well-defined:

$$\forall x \exists y \exists z \forall t A(x,y,z,t).$$

If T^1_2 were valid in $K(F)$, this formula would be too. Assume that it is so. By Lemma 3.3.3, for any standard $\epsilon > 0$, there are $\beta, \beta' \in F$ such that

$$\mu([\![\forall t A(\mathrm{id}_\Omega, \beta, \beta', t)]\!]) \geq 1-\epsilon.$$

Lemma 2.2.2, as F is closed under counter-example functions, then implies the inequality from the claim.

The theorem now follows from the claim, using the hypothesis that there exists f and $\epsilon > 0$ such that the inequality from the claim cannot hold for any β. □

We remark that predicates in $P^N[O(\log n)]$ are exactly those in L^{NP} (logspace with an NP-oracle); see Buss and Hay [17] and Wagner [107], or Krajíček[56].

24.3 PV versus S^1_2

Finally we turn to statement (1).

Theorem 24.3.1 *Assume that no polynomial-size circuit can find a satisfying assignment for more than a $(1-\delta)$ fraction of satisfiable formulas of size k, for some $\delta > 0$ and all $k > 0$ large enough.*

Then PV *(even* $\mathrm{Th}_{\forall_1}(L_{PV})$*) does not prove* S^1_2.

Proof: Let $\mathrm{Sat}(x,y)$ be an open L_{PV} formula formalizing that *x is a propositional formula and y is a satisfying assignment for x.*

We will use it only for formulas x of size n (i.e. encoded by size n strings), and we think of truth assignments y as being encoded by strings of size n too, but we suppress the parameter n in the formulas.

The sample space Ω of a structure to be defined shortly is the set of all n-tuples

$$\omega = (\omega_1, \ldots, \omega_n)$$

where all ω_i are different satisfiable formulas of length n. In particular, samples are n^2-tuples of bits.

Before we define the family F of random variables of the model we need to define a certain formula and a specific counter-example function. Let $f(x,y)$ be a polynomial-time function whose domain is $(\{0,1\}^n)^n \times (\{0,1\}^n)^n$, i.e. both $x = (x_1, \ldots, x_n)$ and $y = (y_1, \ldots, y_n)$ are n-tuples of strings of length n, and which is defined as:

$$f(x,y) := \max_{0 \le i \le n} \bigwedge_{j \le i} \mathrm{Sat}(x_j, y_j).$$

The sentence Max_f is

$$\forall x \exists y \forall z\, f(x,y) \ge f(x,z)$$

(with x, y and z restricted to strings of size n^2, as above).

Claim 1: *The sentence* Max_f *is provable in* $S^1_2(\mathrm{PV})$.

The sentence is an instance of the Σ^b_1-LENGTH-MAX principle that is available in S^1_2. Alternatively, note that the natural proof by induction uses Σ^b_1-LIND which is known to be equivalent to S^1_2. See Buss [11] or Krajíček[56] for details.

Let $h(x)$ be a counter-example function for formula

$$\forall y\, \neg \mathrm{Sat}(x,y)$$

(again implicitly restricted to size n formulas). That is, $h(x)$ is a satisfying assignment for formula x if any exists at all.

Using h we can define a specific counter-example function h^* for the formula

$$\forall z\, f(x,y) \geq f(x,z).$$

Function $h^*(x,y)$ is computed by a polynomial-time algorithm that calls h as an oracle: on input (x,y) (of size n^2) computes $i := f(x,y)$ and outputs z which is equal to y except that y_{i+1} is replaced by $h(x_{i+1})$.

Now we are ready to define F: it is the smallest class of functions defined on Ω closed under composition that contains all PV functions and $h^*(\mathrm{id}_\Omega, y)$. In particular, F is L_{PV}-closed and closed under h^*. Hence PV is $K(F)$-valid (by Lemma 1.4.2).

Note that any function in F is defined by a standard term of $L_{PV} \cup \{h^*(\mathrm{id}_\Omega, y)\}$ and any such term contains at most a standard number of occurrences of h^*.

To show that $S^1_2(\mathrm{PV})$ is not valid in $K(F)$, and hence PV does not prove $S^1_2(\mathrm{PV})$, it suffices to demonstrate that the sentence Max_f is not $K(F)$-valid. This follows from the next claim, using the hypothesis of the theorem.

Witnessing claim 2: *Assume that the sentence* Max_f *is* $K(F)$*-valid. Then for any standard* $\delta > 0$ *there is a circuit of size* n^e*, some standard* $e \in \mathbf{N}$*, that finds a satisfying assignment for at least a fraction* $(1-\delta)$ *of satisfiable formulas of size* n.

To prove the claim assume for the sake of a contradiction that the sentence is $K(F)$-valid. As F is closed under h^*, a counter-example function for formula

$$\forall z\, f(\mathrm{id}_\Omega, y) \geq f(\mathrm{id}_\Omega, z)$$

the assumption together with Lemma 2.2.2 imply that, for an arbitrary standard $\epsilon > 0$, the value of which we shall fix later, there exists element $\beta \in F$ such that

$$\mathrm{Prob}_{\omega \in \Omega}[f(\omega, \beta(\omega)) \text{ is maximal possible }] \geq 1 - \epsilon.$$

Assume that the term defining β contains c calls to $h^*(\mathrm{id}_\Omega, y)$, for some fixed standard $c \geq 1$.

Let Ω_0 and Ω_1 be the sets of $(c+1)$-tuples and $(n-c-1)$-tuples of satisfiable formulas of size n. Hence $\Omega = \Omega_0 \times \Omega_1$. An averaging argument shows that there is an element $b \in \Omega_1$, a tuple of $(n-c-1)$ satisfiable formulas, such that

$$\operatorname*{Prob}_{(\omega_1,\ldots,\omega_{c+1})\in\Omega_0}[f((\omega_1,\ldots,\omega_{c+1},b),\beta((\omega_1,\ldots,\omega_{c+1},b))) \text{ is maximal possible}]$$

$$\geq 1-\epsilon.$$

The rest of the argument is similar (although not identical) to the argument from Krajíček *et al.* [78]; see also Krajíček [56, L.10.2.2]; we give a brief sketch.

The term computing $\beta((\omega_1,\ldots,\omega_{c+1},b))$ asks for $h^*((\omega_1,\ldots,\omega_{c+1},b),y)$, for up to c different values of y. Let V_0 be the set of satisfiable formulas of size n that do not occur among b. For a c-element subset $Q \subseteq V_0$ and a formula $a \in V_0 \setminus Q$ we say that Q *helps* a if for some ordering $a_1,\ldots,a_{c+1}$ of $Q \cup \{a\}$ term β computes a satisfying assignment for a without asking (via h^*) for $h(a)$. In other words, knowing satisfying assignments for formulas from Q and also for all formulas in the tuple b allows us to compute one also for formula a.

There is the complication that term β works correctly for at least a fraction of $(1-\epsilon)$ of all $(c+1)$ tuples of different formulas from V_0 but not for all. Call such a $(c+1)$-tuple *good* if β works correctly for it, and call a formula *a good* if it occurs in at least one good tuple. All good tuples can draw elements only from good formulas. Hence if $1-\delta'$ is the proportion of good formulas in V_0 it must hold that

$$1-\epsilon \leq (1-\delta')^c < 1-\delta',$$

i.e. $\epsilon > \delta'$. Hence taking ϵ to be the δ from Claim 2 will guarantee that there is at least a fraction of $(1-\delta)$ good formulas in V_0. We will construct a polynomial-size circuit computing a satisfying assignment for all good formulas as well as for formulas from b; this is at least a fraction $(1-\delta)$ of all satisfiable formulas.

Let V_1 be the good formulas in V_0. Let m_1 be the size of V_1. There are $\binom{m_1}{c}$ possible sets Q while there are $\binom{m_1}{c+1}$ sets $Q \cup \{a\}$. Hence there is some Q_1 that helps at least

$$\frac{\binom{m_1}{c+1}}{\binom{m_1}{c}} = \frac{(m_1-c)}{c+1}$$

of formulas $a \in V_1$.

Collect into a future advice the $(n-c-1)$ formulas occurring in b, the c formulas from Q_1 and also the satisfying truth assignments for all these $(n-1)$ formulas. This is $(n-1)2n$ bits of advice so far.

Now proceed analogously as before but with set $V_2 := V_1 \setminus Q_1$ in place of V_1. Let m_2 be the cardinality of V_2. Find set $Q_2 \subseteq V_2$ of c formulas that help at least a fraction $(m_2 - c)/(c+1)$ of formulas from V_2. Add this set of c formulas together with some satisfying assignments for them ($2cn$ bits in total) to the advice, and put for the next step $V_3 := V_2 \setminus Q_2$.

The cardinalities of the sets $V_1, V_2, \ldots$ decrease exponentially and in $O(n)$ steps get below some fixed (standard) bound, say below c. Add to the advice also the remaining formulas and some satisfying assignments to them.

The total advice has size $O(n^2)$. The polynomial-time algorithm using this advice (i.e. a polynomial-size circuit) that finds a satisfying assignment for all good formulas and also for formulas in the advice works as follows. Given a, first check if it is one of the formulas in the advice. If so, output the assignment encoded in the advice. Otherwise attempt to find a satisfying assignment to a using term β to all tuples obtained from all possible orderings of sets $Q_i \cup \{a\}$, $i = 1, 2, \ldots$ completed to an n-tuple by b. It follows from the definition of sets Q_i that if a is good, the algorithm succeeds. □

24.4 Transfer principles

The topic of this section does not really fit with the rest of the chapter (or into any other one) and would warrant its own chapter. However, it is a bit peripheral in relation to our main line of investigation and we shall discuss it here just briefly. It may be useful in some future constructions.

Consider a structure $K(F)$ defined by a family of random variables F on some sample space Ω. An element $\alpha \in F$ can be viewed from two different perspectives: as an element of the structure $K(F)$ or as an element of the ambient model $\mathcal{M}$, a function defined on Ω obeying some particular restrictions defining F. Thus properties of elements of $K(F)$ can be translated into properties of (some) elements of $\mathcal{M}$, and vice versa. This observation was first used, I believe, by Takeuti [102] who considered Boolean-valued models of set theory. Below we note one specific example of such a translation to illustrate this phenomenon.

Let us consider structure $K(F_{PV})$, with F_{PV} being the family of polynomial-time functions defined on $\{0,1\}^n$ (see Section 3.2 or Section 24.1 in this Chapter).

Let Q be a propositional proof system in the sense of Cook and Reckhow [31]; see Chapter 27. Let $\alpha_Q \in F_{PV}$ be Q restricted to $\{0,1\}^n$. In particular,

$$[\![\alpha_Q \text{ is a formula }]\!] = 1_{\mathcal{B}}$$

but even more:

$$[\![\text{Taut}(\alpha_Q)]\!] = 1_{\mathcal{B}}$$

where $\text{Taut}(x)$ is a bounded universal formula $\forall y(|y| \leq |x|)\text{Sat}(x,y)$, with $\text{Sat}(x,y)$ being an open L_{PV}-formula defining the relation y is a truth if assignment then it satisfies formula x (as in the proof of Theorem 24.3.1).

To see this note that

$$[\![\text{Taut}(\alpha_Q)]\!] = \bigwedge_{\beta} [\![|\beta| \leq |\alpha_Q| \to \text{Sat}(\alpha_Q, \beta)]\!]$$

and that for any β

$$\text{Prob}_{\omega \in \{0,1\}^n}[\ |\beta(\omega)| \leq |\alpha_Q(\omega)| \to \text{Sat}(\alpha_Q(\omega), \beta(\omega))\ ] \ = \ 1$$

as only tautologies get into the range of Q.

Now let P be another proof system (represented in $K(F_{PV})$ by its symbol in L_{PV}) and assume

$$[\![P \vdash \alpha_Q]\!] = 1_{\mathcal{B}},$$

that is

$$[\![\exists z P(z) = \alpha_Q]\!] = 1_{\mathcal{B}}.$$

Using the properties of witnessing quantifiers it follows that for every standard $\epsilon > 0$ there is a polynomial-time function γ such that

$$\text{Prob}_{\omega}[\ P(\gamma(\omega)) = Q(\omega)\ ] \ > \ 1 - \epsilon.$$

In other words, γ translates Q-proofs into P-proofs of the same formula, i.e. it is a polynomial simulation of Q by P (albeit with an error ϵ). Let us call such a function an **approximate p-simulation**, leaving the quantification via ϵ out of this discussion (the parameter ϵ could be avoided if we used a compact family, e.g. functions computable in subexponential time, in place of F_{PV}; see Section 3.5).

A function $\sigma \in F_{PV}$ may satisfy the equation

$$[\![\text{Taut}(\sigma)]\!] = 1_{\mathcal{B}}$$

even if σ is not a proof system in the Cook and Reckhow sense. Such a σ may make errors but these should be hard to detect. In particular, it must hold that for all polynomial time β

$$\text{Prob}_{\omega}[\ |\beta(\omega)| \leq |\sigma(\omega)| \to \text{Sat}(\sigma(\omega), \beta(\omega))\ ] \ > \ 1 - \epsilon$$

for all standard $\epsilon > 0$. A polynomial-time function σ satisfying this condition may be called a **pseudo proof system**.

Thus properties of formulas and their proofs can be translated, via $K(F_{PV})$, into properties of pseudo proof systems and their approximate p-simulations, and vice versa.

25

Pseudorandom sets and a Löwenheim–Skolem phenomenon

Let $K(F)$ be an L-structure with Ω its sample space. There seems to be a natural concept of a substructure of $K(F)$: a structure $K(F_0)$ with the same sample space but with possibly a smaller, L-closed family of random variables $F_0 \subseteq F$. As the sample space of both structures is the same, so is the Boolean algebra $\mathcal{B}$, and the atomic sentences using parameters from F_0 get the same truth-values. This mirrors well the classical concept of a substructure.

There is, however, another way to create structures that are in a good sense substructures of $K(F)$. Instead of taking a smaller family of random variables, take a smaller sample space $\Omega_0 \subseteq \Omega$ (we need to maintain the general requirement for the method that Ω_0 is infinite) but the same family F.

Let us use the following notation: $\mathcal{A}_0$, $\mathcal{B}_0$ are the resulting Boolean algebras, $[\![\ldots]\!]_0$ the truth values in $\mathcal{B}_0$, and μ_0 will be the (standard) measure on $\mathcal{B}_0$. As the Boolean algebras of truth values are different in each case we cannot compare the truth values of $L(F)$-sentences in the two structures. But we can compare their measures.

Assume the space Ω_0 is chosen in such a way that:

(*) *For any atomic formula $A(x_1,\ldots,x_k)$ and any $\alpha_1,\ldots,\alpha_k \in F$ it holds that:*

$$\mu([\![A(\alpha_1,\ldots,\alpha_k)]\!]) = \mu_0([\![A(\alpha_1,\ldots,\alpha_k)]\!]_0).$$

We will say that language L is closed under definition by cases by open formulas if every open L-formula is equivalent in $\mathbf{N}$ (recall that always $L \subseteq L_{all}$) to an atomic one. Such an equivalence is a universal statement and hence will be valid in any L-closed structure by Lemma 1.4.2.

In the next lemma we use the notation just introduced.

Lemma 25.1 *Assume L is closed under definition by cases by open formulas. Assume $\Omega_0 \subseteq \Omega$, $\Omega_0 \in \mathcal{M}$ and is infinite, and F is an L-closed family. Assume also that the hypothesis $(*)$ above is satisfied.*

Then for any $L(F)$-sentence A it holds that:

$$\mu([\![A]\!]) = \mu_0([\![A]\!]_0).$$

In particular,

$$[\![A]\!] = 1_{\mathcal{B}} \text{ iff } [\![A]\!]_0 = 1_{\mathcal{B}_0}.$$

Proof: The statement is proved by induction on the logical complexity of the sentence. For atomic sentences this is the hypothesis $(*)$ and for open sentences this follows from the assumption that L is closed under definition by cases by open formulas, i.e. any open formula is equivalent to an atomic one.

Now assume A has the form $\exists x B(x)$, where B is an $L(F)$-formula. Family F is closed under definition by cases by open formulas (this follows from the corresponding assumption about L) and hence by Lemma 3.3.2 the following holds: for any $k \in \mathbf{N}$ there is $\alpha_k \in F$ such that

$$\mu([\![A]\!]) \geq \mu([\![B(\alpha_k)]\!]) > \mu([\![A]\!]) - \frac{1}{k}.$$

Hence $\mu([\![A]\!])$ is the limit of values $\mu([\![B(\alpha_k)]\!])$. By induction assumption

$$\mu([\![B(\alpha_k)]\!]) = \mu_0([\![B(\alpha_k)]\!]_0).$$

It follows that $\mu([\![A]\!]) \leq \mu_0([\![A]\!]_0)$. Identically one proves the opposite inequality.

The universal quantifier is treated analogously. □

Now we link these considerations with the concept of a pseudorandom distribution from computational complexity theory. We will restrict the discussion to language L_{PV}, sample space $\Omega = \{0,1\}^n$ and the family F_{PV} of random variables computed by (standard) polynomial-time algorithms.

However, we need to consider a slightly more general form of Ω_0; instead of mere subsets $\{0,1\}^n$ we need to allow Ω_0 to be a probabilistic distribution on $\{0,1\}^n$. Such a distribution assigns to elements of $\{0,1\}^n$ arbitrary probabilities (defined in $\mathcal{M}$) summing up to 1 on the whole space. In defining $\mathcal{B}_0$ from $\mathcal{A}_0$ the counting measure is replaced by the new measure. In the text below we assume this more general definition of μ_0 and $[\![\ldots]\!]_0$.

The following two definitions are weaker versions of the standard definitions; see Goldreich [36]. The weakening is that for our purposes it suffices to consider indistinguishability with respect to the uniform polynomial-time algorithm while in the standard definitions one needs to allow randomized polynomial-time algorithms or even non-uniform polynomial-time algorithms (i.e. polynomial-size circuits).

Definition 25.2 Families of probabilistic distributions $(\Delta_k)_k$ and $(\Gamma_k)_k$ on $\{0,1\}^k$, $k \in \mathbf{N}$, are called **computationally indistinguishable**, denoted

$$(\Delta_k)_k \equiv_c (\Gamma_k)_k,$$

if the following condition holds.

For any unary predicate $P \subseteq \{0,1\}^*$ that is computable in polynomial time and for any $c \in \mathbf{N}$ we have:

$$|\ \mathrm{Prob}_{x \in \Delta_k}[x \in P] \ - \ \mathrm{Prob}_{y \in \Gamma_k}[y \in P]\ | \ < \ k^{-c}$$

for all $k \in \mathbf{N}$ large enough.

The probabilities in the definition are computed according to the respective distributions.

Definition 25.3 The family of probabilistic distributions $(\Delta_k)_k \subseteq \{0,1\}^k, k \in \mathbf{N}$, is called **pseudorandom** iff

$$(\Delta_k)_k \equiv_c (\{0,1\}^k)_k,$$

where $(\{0,1\}^k)_k$ denotes the family of uniform distributions.

Lemma 25.4 *Assume $(\Delta_k)_k \subseteq \{0,1\}^k$ is a pseudorandom family of probabilistic distributions. Take for Ω_0 the distribution Δ_n on $\{0,1\}^n$, and $\Omega := \{0,1\}^n$ (i.e. the uniform distribution). Let $L = L_{\mathrm{PV}}$ and $F = F_{\mathrm{PV}}$.*

Then (using the notation above) for all $L_{\mathrm{PV}}(F_{\mathrm{PV}})$-sentences A it holds that

$$\mu([\![A]\!]) \ = \ \mu_0([\![A]\!]_0).$$

In particular,

$$[\![A]\!] = 1_{\mathcal{B}} \ \mathit{iff}\ [\![A]\!]_0 = 1_{\mathcal{B}_0}.$$

Proof: The lemma follows from Lemma 25.1 (for distributions) once we verify the hypothesis $(*)$. To do this note that any atomic $L_{PV}(F_{PV})$ sentence $A(\alpha_1,\ldots,\alpha_k)$ defines a polynomial-time predicate on Ω: $\omega \in \langle\!\langle A(\alpha_1,\ldots,\alpha_k)\rangle\!\rangle$.

By the definition of pseudorandomness it follows that the counting measure and the measure in the sense of Δ_n of

$$\langle\!\langle A(\alpha_1,\ldots,\alpha_k)\rangle\!\rangle$$

differ at most infinitesimally. Hence

$$\mu([\![A(\alpha_1,\ldots,\alpha_k)]\!]) = \mu_0([\![A(\alpha_1,\ldots,\alpha_k)]\!]_0).$$

□

It is not known how to define a pseudorandom distribution efficiently, in a sense to be described shortly. But we note a simple consequence of Chernoff's inequality, using the fact that the set of standard polynomial-time algorithms is included in an $\mathcal{M}$-definable family of algorithms of arbitrarily small non-standard cardinality.

Lemma 25.5 *For any $s > \mathbf{N}$ there is an infinite distribution $\Omega_0 \subseteq \{0,1\}^n$, $\Omega_0 \in \mathcal{M}$, of cardinality*

$$|\Omega_0| \leq n^s$$

and such that for any $P \subseteq \{0,1\}^n$ definable by a (standard) polynomial-time algorithm and for any $c \in \mathbf{N}$ we have:

$$|\ \mathrm{Prob}_{x\in\Omega_0}[x \in P]\ -\ \mathrm{Prob}_{y\in\{0,1\}^n}[y \in P]\ |\ <\ n^{-c}.$$

The set Ω_0 defines a structure and satisfies the properties stated in Lemma 25.4. But note that Lemma 25.5 does not suggest how to define Ω_0 efficiently, it only asserts its existence.

There is a conjectural way to find such a distribution by listing its elements using a polynomial-time algorithm or, more precisely, by sampling the distribution in polynomial time. This uses a now accepted hypothesis (some even say 'standard' hypothesis) about the existence of a pseudorandom number generator (or, equivalently, a one-way function). This conjecture is crucial for the foundation of cryptography based on computational complexity; see Goldreich [36].

We will formulate the next definition only for $\{0,1\}^n$. Note that because we talk again only about uniform algorithms the next definition is weaker than the standard one too.

Definition 25.6 A polynomial-time function $g : \{0,1\}^m \to \{0,1\}^n$ is a **pseudorandom number generator** if $m < n$ and the distribution assigning to $\omega \in \{0,1\}^n$ the probability

$$\frac{|g^{(-1)}(\omega)|}{2^m}$$

is a pseudorandom distribution on $\{0,1\}^n$. In other words,

$$|\ \mathrm{Prob}_{x\in\{0,1\}^m}[g(x) \in P]\ -\ \mathrm{Prob}_{y\in\{0,1\}^n}[y \in P]\ |\ <\ n^{-c}$$

holds for any polynomial-time predicate P on $\{0,1\}^n$ and any $c \in \mathbf{N}$.

A pseudorandom generator g defines a distribution on $\{0,1\}^n$ that can be sampled as follows: pick $u \in \{0,1\}^m$ uniformly at random and take sample $\omega := g(u)$. This can be used to define a structure $K(F_g)$ as follows:

- The sample space is $\{0,1\}^m$.
- The family F_g consists of functions $\beta : \{0,1\}^m \to \mathcal{M}_n$ that have the form

$$\beta = \alpha \circ g$$

  for some $\alpha \in F_{\mathrm{PV}}$.

It is then clear that this structure has the same properties as the structure from Lemma 25.4. We state just the latter property.

Lemma 25.7 *Assume g is a pseudorandom generator. Then any L_{PV} sentence A is valid in $K(F_{\mathrm{PV}})$ iff it is valid in $K(F_g)$.*

Pseudorandom generators are conjectured to exist for any $m \leq n^\epsilon$, with arbitrary standard $\epsilon > 0$. The sample space of $K(F_g)$ then has size at most 2^{n^ϵ}, i.e. at most

$$2^{(\log|\Omega|)^\epsilon}.$$

In the non-uniform construction of sample space Ω_0 in Lemma 25.5 the size is even much smaller, less then

$$(\log|\Omega|)^s,$$

for arbitrarily small but non-standard s. It thus seems coherent to interpret these constructions as a Löwenheim–Skolem type phenomenon.

26

Sampling with oracles

In this chapter we define two fairly simple models of PV, a first-order $K(F_{\text{oracle}})$ and its second-order expansion $K(F_{\text{oracle}}, G_{\text{oracle}})$. The latter structure allows for a natural interpretation of structural complexity results about a random oracle.

26.1 Structures $K(F_{\text{oracle}})$ and $K(F_{\text{oracle}}, G_{\text{oracle}})$

Take a non-standard $n > \mathbf{N}$ and put $h := 2^{n^n}$ (we only need that h is bigger than any 2^{n^c}, for any $c \in \mathbf{N}$). The sample space is

$$\Omega_{\text{oracle}} := \{0,1\}^h.$$

The family of random variables F_{oracle} consists of all functions

$$\alpha : \Omega_{\text{oracle}} \to \mathcal{M}$$

such that there is a (standard) polynomial-time oracle machine M^Q such that for all samples $\omega \in \Omega_{\text{oracle}}$:

$$\alpha(\omega) := M^{\omega}(1^{(n)}).$$

As the machine is polynomial time any potential oracle query has length bounded above by n^c, for some $c \in \mathbf{N}$, and hence all samples ω contain answers to all such queries.

Lemma 26.1.1 *F_{oracle} is L_{PV}-closed and closed under definitions by cases by open $L_{\text{PV}}(F_{\text{oracle}})$ formulas.*

An immediate corollary of the lemma and Lemma 1.4.2 is this.

Corollary 26.1.2 $\mathrm{Th}_\forall(L_{PV})$ *is valid in* $K(F_{\mathrm{oracle}})$. *In particular,* PV *is valid in the structure.*

Now we will put a second-order structure, a family of functions G_{oracle}, on top of $K(F_{\mathrm{oracle}})$.

An element $\Theta \in G_{\mathrm{oracle}}$ is a map

$$\Theta : F_{\mathrm{oracle}} \to F_{\mathrm{oracle}}$$

determined by a polynomial-time oracle machine N^Q and defined by

$$\Theta(\alpha)(\omega) := N^\omega(\alpha(\omega)).$$

It is clear that $\Theta(\alpha)$ is computed (on input $1^{(n)}$) by the composition of machine M^Q computing α with the machine N^Q.

Note that any L_{PV} function symbol has a natural interpretation in G_{oracle}: a PV function f computed by a polynomial-time machine N is represented by the element of G_{oracle} determined by N (which asks no oracle queries). In this sense the following holds.

Lemma 26.1.3 *There is a family* $G_{PV} \subseteq G_{\mathrm{oracle}}$ *such that* PV *is valid in the structure* $K(F_{\mathrm{oracle}}, G_{PV})$.

26.2 An interpretation of random oracle results

We will demonstrate in this section that structural complexity results of the form that something happens in the relativized world with probability 1 for a random oracle can be naturally interpreted in $K(F_{\mathrm{oracle}}, G_{\mathrm{oracle}})$.

One can imagine a random oracle as a subset of $\mathbf{N}$ to which any number belongs with probability $\frac{1}{2}$. This is sound if one studies (as is usually the case) the behavior of an algorithm of a limited time or space complexity on inputs of a bounded size (and hence the size of oracle queries is bounded too). This view of random oracles obviously generalizes to viewing a random oracle as a random element of $\{0,1\}^m$ for a non-standard m. A more general approach considers measurable sets of oracles (i.e. subsets of the power set of $\mathbf{N}$) and the results talk about measures of sets of oracles having a certain property. This actually also relates to viewing random oracles as a random element of $\{0,1\}^m$ as the counting measure is a non-standard refinement of the Lebesgue measure. We will skip details of these comments and appeal just to the intuition of the reader.

A typical result in this area is the following.

Theorem 26.2.1 (Bennett–Gill [9]) *For a random oracle $A \subseteq \mathbf{N}$, with probability* 1 *it holds that:*

$$P^A \neq NP^A.$$

Let $R^Q(x,y)$ be a binary relation defined by a polynomial-time oracle machine $N^Q(x,y)$. Assume for simplicity of formulas that there is a constant $c \in \mathbf{N}$ such that for any oracle A

$$R^A(x,y) \rightarrow |y| \leq |x|^c.$$

Suppose the relation R^Q witnesses Theorem 26.2.1; that is, no polynomial-time oracle machine M^Q satisfies

$$\forall x\, (\exists y N^A(x,y)) \rightarrow N^A(x, M^A(x))$$

with a non-zero probability for a random oracle $A \subseteq \mathbf{N}$.

Take $\Theta \in G_{\text{oracle}}$ determined by N^Q. The random oracle result has the following interpretation in $K(F_{\text{oracle}}, G_{\text{oracle}})$.

Theorem 26.2.2 *The open formula $\Theta(x,y) = 1$ has no Skolem function in $K(F_{\text{oracle}}, G_{\text{oracle}})$. In particular, for all $\Gamma \in G_{\text{oracle}}$*

$$[\![\forall x, y\, (\Theta(x,y) = 1 \rightarrow \Theta(x, \Gamma(x)) = 1)]\!] = 0_{\mathcal{B}}.$$

We remark that it is natural to define a non-uniform version of the structure $K(F_{\text{oracle}}, G_{\text{oracle}})$ where the role of polynomial-time oracle machines is taken by decision trees (as in $K(F_{\text{rud}}, G_{\text{rud}})$) of depth $\leq n^c$, for arbitrary $c \in \mathbf{N}$.

It is not without interest to note that this non-uniform version can be seen as a 'short' version of $K(F_{\text{rud}}, G_{\text{rud}})$. Namely, if we denote the length of samples in both cases by ℓ then the depth of the trees as well as the length of the values used in $K(F_{\text{rud}}, G_{\text{rud}})$ is $\ell^{o(1)}$ while in the non-uniform version of $K(F_{\text{oracle}}, G_{\text{oracle}})$ it is only $(\log\log \ell)^{O(1)}$. (The double log is because of our generous bound 2^{n^n} in the definition of the sample space but 2^{n^t} would suffice for any non-standard t and hence the estimate should be more like $(\log \ell)^{o(1)}$.)

PART VIII

Proof complexity of EF and beyond

27

Fundamental problems in proof complexity

In this last part of the book we shall consider strong proof systems. We start by recalling in this chapter briefly a very general background of proof complexity, and by spelling out a few open problems that we consider fundamental for further development. This exposition is included in order to give the reader some idea of what motivates the choice of topics we study in later chapters. Details can be found in the references given in the Introduction (or in the text below).

Cook and Reckhow [31] defined a general **proof system** for propositional logic to be a polynomial-time computable function P defined on $\{0,1\}^*$ whose range is exactly the set TAUT of propositional tautologies in the DeMorgan language $0, 1, \wedge, \vee, \neg$ (see Chapter 17). Any string that P maps to a tautology τ is called a **P-proof** of τ. This definition subsumes the usual calculi for propositional logic that one encounters in textbooks and, in particular, those defined in Chapter 17: given such a calculus interpret it as a function that maps a string that is a valid proof to the formula being proved, and all other strings to some fixed tautology (for example, to 1). Such a function will satisfy the Cook–Reckhow definition as in all logical calculi it is recognizable by a p-time algorithm whether a string is a valid proof or not. Note that the fact that the range of the function constructed in this way from a propositional calculus is exactly the set TAUT is equivalent to the soundness and the completeness of the calculus.

The main question about proof systems is whether there exists one in which all tautologies have a proof of size polynomially bounded in terms of the size of the tautology; such a proof system is called **p-bounded** (in some original texts it was called *super*). Cook and Reckhow [31] observed that the existence of a p-bounded proof system is equivalent to the statement that the computational complexity class NP is closed under complementation, i.e. that NP = coNP. It is widely believed that NP $\neq$ coNP (although perhaps not as widely believed as P $\neq$ NP) and hence that for every proof system P there is an infinite set of tautologies whose shortest P-proofs cannot be bounded by a polynomial in their length (in fact, it is expected that such hard tautologies can be found for which the shortest P-proofs must have exponential size). In principle thus one can show that NP is not closed under complementation, and consequently also that P $\neq$ NP, by proving sufficiently strong and general lengths-of-proofs lower bounds.[1] The main problem of proof complexity is therefore:

Show that no p-bounded proof system exists (i.e. that NP $\neq$ coNP).

There are many proof systems utilizing ideas from logic, combinatorics, algebra or geometry, and superpolynomial lower bounds are known for only a few of them (see Krajíček [56] or surveys by Pudlák [89] or Krajíček[68]). The best-understood among them is resolution, which is of independent interest for automated theorem proving. Problems to demonstrate lower bounds for stronger proof systems, based on more complicated mathematical ideas, have led to increasingly complex combinatorial problems. At present we cannot even rule out that a Frege proof system (see Definition 17.1.2) is p-bounded.

A particular for place among lower bound methods is held by feasible interpolation. It is a method that arguably applies to the most varied class of proof systems. Here one shows that from a short proof of the disjointness of two NP sets (restricted to $\{0,1\}^n$, $n \geq 1$) it is possible to extract a feasible algorithm that separates them. Shorter proofs lead to more efficient algorithms. Hence pairs of disjoint NP sets that are hard to separate yield formulas that need long proofs. Computational complexity theory does not offer such examples unconditionally (suitable examples follow from the conjectured existence of one-way functions) but sometimes there is a 'monotone' version of the argument that needs only lower bounds for monotone Boolean circuits (or other 'monotone' computational models), and such lower bounds are sometimes known. This method, and its variants, applies to resolution, cutting planes proof systems, algebraic and geometric proof systems, and recently also to the OBDD proof system. However, various cryptographic assumptions (e.g. the security of the RSA) are known to imply that the method does not apply to strong proof systems (but see Krajíček [71]; references to other papers on feasible interpolation can also be found there).

It seems therefore that combinatorics alone is inadequate to tackle strong proof systems. This may be reminiscent of Boolean complexity: the combinatorial problems arising there are very complicated and this hindered progress

[1] This approach to the P vs. NP problem is sometimes called Cook's program.

severely during the last twenty years. But proof complexity is fortunate to have another – genuinely non-combinatorial – facet: links with bounded arithmetic theories. These theories capture the informal concept of 'feasible' reasoning in a way analogous to how polynomial-time algorithms (and other restricted classes of circuits like AC^0, TC^0, etc.) capture the notion of a feasible algorithm. One can translate between proofs in these weak first-order systems and short proofs in commonly studied propositional proof systems; one can be thought of as a uniform version of the other. The upshot is that one can think of lengths-of-proofs lower bounds as about problems of how to construct suitable models of particular bounded arithmetic. This is what we formulated in Part V (Chapter 18).[2]

The existence of a p-bounded proof system (i.e. the NP vs. coNP problem) appears to be a very hard problem and it makes a good sense to consider also (presumably) easier problems which nevertheless retain the spirit of the main problem. The following three problems seem to me to be particularly stimulating:

(1) Prove the conjecture that NP $\neq$ coNP from a plausible computational hardness assumption.
(2) Prove a super-polynomial lengths-of-proofs lower bound for the Extended Frege system EF.
(3) Decide whether or not there exists a p-optimal (or an optimal) proof system.

Concerning Problem (1) recall that there are several fundamental conjectures in computational complexity theory that do follow from computational hardness assumptions about Boolean circuits. These assumptions have a rather similar form:

- *Every Boolean circuit performing a particular, explicitly given, computational task has to be large.*

These include the conjectures that P $\neq$ NP, P $=$ BPP (i.e. that the randomization does not increase the power of p-time algorithms), and the conjecture that a strong pseudorandom generator that is useful for cryptography exists. In particular, using the trivial inclusion of P $\subseteq$ P/poly and the NP-completeness of SAT we can reduce P $\neq$ NP to the hardness assumption that

- SAT cannot be decided by polynomial-size Boolean circuits.

[2] There are also many direct connections of bounded arithmetic with algorithmic complexity; see Krajíček [56] or Cook and Nguyen [30].

Using Nisan–Wigderson generators (see Section 29.4) Impagliazzo and Wigderson [48] proved that P = BPP assuming that

- Boolean circuits computing a complete language in the exponential time class E (time bound $2^{O(n)}$) must have size at least $2^{\Omega(n)}$.

And finally, Hastad *et al.* [41] proved that strong pseudorandom generators exists, assuming that a one-way function exists. In particular, we have a reduction of the existence of a strong pseudorandom generator to the following concrete hardness assumption:

- Factoring is intractable by polynomial-size Boolean circuits. In particular, polynomial-size circuits can factor only a subpolynomial fraction of integers that are products of two primes of comparable sizes.

It would thus be quite interesting (and mathematically appealing) to have also a reduction of the conjecture that NP $\neq$ coNP to an assumption of a similar form. See Krajíček[67] for more discussion of this topic.

Problem (2) is significant for a variety of reasons (see Krajíček [53]). The extended Frege system is considered to be a pivotal case in proof complexity research should we progress from proving lower bounds for (seemingly) weak systems to attack strong systems. There is also another reason to do with the link between EF and Cook's theory PV (see Chapter 23 for PV). It is known that the consistency of P $\neq$ NP with PV would follow from *any* super-polynomial lower bound for Extended Frege proofs. Such a consistency would be a breakthrough as a large part of contemporary complexity theory can be formalized in PV (or in its mild extension).

Problem (3) raises quite a fundamental issue, interesting in its own right. If we cannot resolve the existence of a p-bounded proof system the next natural question is whether there exists a proof system that is in a sense the strongest, as good as any other proof system. The right (quasi)ordering of proof systems, called **p-simulation**, has been defined already in Cook and Reckhow [31]: proof system P p-simulates Q iff there exists a p-time function f such that for all $w \in \{0,1\}^*$ we have $P(f(w)) = Q(w)$. In other words, simulation f translates Q-proofs into P-proofs of the same formulas. As f is p-time it has also polynomial growth: the length of $f(w)$ is at most polynomially longer than that of w. If we keep the requirement of the polynomial growth but drop the p-time computability of f we talk of **simulation**. We say that P is **p-optimal** (resp. optimal) iff it p-simulates (resp. simulates) any other proof system Q. Starting with Krajíček and Pudlák [73] this question (and its non-uniform version) has been studied from many angles, and links with various other topics in complexity theory and mathematical logic have been found (see Krajíček [56]). In particular, the non-existence of a p-optimal proof system implies that EXP $\neq$ NEXP (exponential time with $2^{n^{O(1)}}$ bound) and hence also P $\neq$ NP, and Krajíček [64] has proved that at least one of the following three conjectures is true:

- There is a function $f : \{0,1\}^* \rightarrow \{0,1\}$ computable in E that has circuit complexity $2^{\Omega(n)}$.
- NP $\neq$ coNP.
- There is no p-optimal propositional proof system.

There is a problem about proof search closely related to Problem (3), namely:

(3′) Is there an optimal way of searching for proofs of unsatisfiability?

As mentioned earlier, proof complexity of resolution is related to automated theorem proving. In fact, known algorithms essentially search for a proof in resolution or in some of its variants (e.g. tree-like resolution). It is quite an interesting question whether one can define an algorithm searching for proofs in a stronger proof system that is 'significantly better' than any algorithm yielding resolution proofs. In any case, searching for proofs is a classical topic, connected with an early history of computational complexity: a version of the P vs NP problem is a question posed by Gödel [35] about searching for proofs. There is also a formal link between Problems (3) and (3′): Krajíček and Pudlák [73] proved that a p-optimal proof system exists iff there exists a deterministic algorithm recognizing TAUT that has at most polynomial slow-down over any other such algorithm on inputs from TAUT.

Problem (3) has several other interesting facets. It has deep links with quantitative versions of Gödel's theorem (see Krajíček and Pudlák [73]) and with NP search problems. Interestingly, Cook and Krajíček [29] defined a concept of a proof system with advice and showed that there is an optimal system in that class.

All this discussion aims at justifying the following problem as a worthwhile target of research:

Prove a super-polynomial lower bound for EF from a plausible computational hardness hypothesis.

This is clearly a common weakening of Problems (1) and (2) and, in fact, there are also links (via proof complexity generators; see Chapter 29) to Problem (3).

Our strategy for an attack on this problem is clear: using an approach analogous to that of Chapter 18 we want to find a structure (formed by a family of random variables) in which the soundness of EF is valid while some tautology is not. These two properties of the model are de facto properties of the family of random variables forming the model, and the statement that the family indeed has the properties will be the hypothesis to which the lengths-of-proofs lower bound for EF will reduce. Of course, the all-important qualification *plausible* hypothesis should not slip our mind.

Let us remark that in this respect forcing with random variables relates to feasible interpolation mentioned earlier, where a lower bound is also reduced to a computational hypothesis (see the earlier discussion).

28

Theories for EF and stronger proof systems

In this chapter we recall the correspondence of EF with some bounded arithmetic theories. We shall consider both first-order and second-order theories as both possibilities may be convenient for different constructions.

28.1 First-order context for EF: PV and S^1_2

For the definitions of theories PV and S^1_2 see Chapter 23. The relation between EF and PV discovered by Cook [28] was, in fact, the first example of a general correspondence between proof systems and bounded arithmetic theories (see Krajíček and Pudlák [73]). In the following we briefly summarize the relation for the case of EF and PV; the same relation of EF to S^1_2 follows from the $\forall\Sigma^b_1$-conservativity of $S^1_2(\mathrm{PV})$ over PV proved by Buss [11]; see also Section 23.2.

Let $A(x)$ be a Π^b_1-definition (see Section 23.2) in language L_{PV} of a coNP-set of numbers. Assume $A(x)$ has the form $\forall y(|y| \leq |x|^k \rightarrow B(x,y))$ with $B(x,y)$ an open L_{PV}-formula (defining thus a polynomial-time predicate).

Fix length n to bound $|x|$ and construct a propositional formula $||A(x)||^n$ as in the proof of the NP-completeness of satisfiability: the formula has n atoms $p_1,\ldots,p_n$ for bits of an x, $m = n^k$ atoms $q_1,\ldots,q_m$ for bits of a potential y, and also atoms $r_1,\ldots,r_s$ for $s = n^{O(1)}$ bits of values on nodes of a fixed circuit C_n computing from $\overline{p}, \overline{q}$ the truth value of predicate $B(x,y)$. Formula $||A(x)||^n$ says, in a DNF form, that if $\overline{r}$ are correctly computed by circuit C_n from inputs $\overline{p}, \overline{q}$ then the output of the computation is 1. Having any b of length n with bits $b(1),\ldots,b(n)$ denote by $||A(x)||^n(b)$ the propositional formula with $b(i)$ substituted for p_i, and with the remaining atoms $\overline{q}$ and $\overline{r}$ left unsubstituted. Clearly then b satisfies $A(x)$ iff $||A(x)||^n(b)$ is a tautology.

The relation between EF and PV is as follows:

(1) If PV proves $\forall x A(x)$ then tautologies $||A(x)||^n(b)$ have polynomial-size EF-proofs (and these are definable from $1^{(n)}$ in PV).
(2) PV proves the soundness of EF and, for any another proof system Q, if PV proves also the soundness of Q then EF p-simulatesQ (and the simulation is definable in PV).

Crucially, the first property has a form of a converse that underlines our approach to lower bounds (Chapter 18). We shall formulate this in some detail in the next section in the second-order formalism (as this is the one used in Chapter 18).

28.2 Second-order context for EF: VPV and V_1^1

The second-order theory V_1^1 has been defined in Section 23.3. Theory VPV (defined in Cook and Nguyen [30]) is also second-order and has an analogous relation to V_1^1 as PV has to S_2^1 (this goes under the name RSUV isomorphism; see Krajíček [56] or Cook and Nguyen [30] or the model-theoretic explanation in Section 23.3).

The first-order part of the language of theory VPV is L_{PA}. The second-order part includes symbols for all polynomial-time functionals $f(\overline{x}, \overline{X})$ whose arguments are both numbers and bounded sets (i.e. strings) and values can be either numbers or strings too. The restriction *polynomial time* means here polynomial in the length of the string inputs. The axioms for VPV include definitions for all of these functions based on Cobham's theorem, quite analogous to the definition of PV in Section 23.1. VPV proves (as PV does) the induction scheme for open formulas. It is a universal theory.

The property that is a form of converse to (1) from the previous section is provided by Theorem 18.3.1. But we want to generalize it a bit. Namely, in Chapter 18 we have considered only constant-depth formulas (denoted T_k there); here we want to allow any formulas or, in fact, circuits. The reason for considering only constant-depth formulas there was that in the theories corresponding to constant-depth Frege systems one cannot prove that an arbitrary formula can be evaluated on an arbitrary truth assignment. In other words, in models of such theories we cannot talk about the validity of general formulas. In VPV and V_1^1 any circuit can be evaluated (and in U_1^1 any formula). Hence we may allow T_k to be arbitrary circuits. By Fagin's theorem the satisfiability of circuits can be defined by both $\Sigma_1^{1,b}$- and $\Pi_1^{1,b}$-formulas. Let Sat denote in the next theorem a

$\Sigma_1^{1,b}$-definition of satisfiability for which the basic properties can be proved in VPV (see Krajíček [56]).

The theorem stated for EF and VPV and generalized as just explained is as follows.

Theorem 28.2.1 *Let T_k be circuits of size $k^{O(1)}$, for all $k \in \mathbf{N}$. Assume that for an arbitrary choice of non-standard n there is a structure $K(F,G)$ (built over $\mathcal{M}_n$) satisfying the following three conditions:*

(1) $K(F,G)$ is L_n-closed.
(2) VPV is valid in $K(F,G)$.
(3) There is a truth assignment $\Gamma \in G$ to atoms of T_n that falsifies the circuit in $K(F,G)$, i.e. $\mathrm{Sat}(n,\Gamma,\neg T_n)$ is valid.

Then there is a standard $\epsilon > 0$ such that for no $k \in \mathbf{N}$ large enough has T_k an EF-proof of size less than 2^{k^ϵ}.

The proof is the same as that of Theorem 18.3.1 (outlined in Chapter 18), using the fact that VPV proves the soundness of EF.

28.3 Stronger proof systems

By a *strong* proof system different people mean different things. Sometimes it is simply meant to be a proof system for which we are unable, at the present time, to demonstrate that it is not p-bounded. In this respect there are countless strong systems. But that appears to be a rather ad hoc criterion.

The choice of EF as the first example of a strong proof system is motivated by several considerations, none of which is really convincing on its own but which together seem to make a good case for EF.

First, many weaker proof systems operate not with arbitrary circuits but with some restricted class of circuits. It may thus happen that an eventual lower bound proof for the weaker system will somehow utilize the circuit restriction. EF on the other hand is p-equivalent to a proof system Circuit Frege CF, a kind of a Frege system operating with unrestricted circuits.[1] Thus a lower bound for EF cannot use any specific circuit properties.

Second, while we do have several particular proof systems and several classes of proof systems that we think may be stronger than EF (i.e. may p-simulate EF but not be p-simulated by it), they can all be p-simulated by a 'simple' extension

[1] This is often claimed as folklore but the definition of CF, a Frege system for circuits, is not entirely straightforward and has been properly done only recently by Jeřábek [50].

of EF by a p-time set of extra axioms. The particular proof systems possibly stronger than EF include WF of Jeřábek [50] (see also Section 30.1), a proof system corresponding to S^1_2 extended by the dual weak PHP for p-time functions, or an algebraic proof system UENS (the abbreviation stands for 'unstructured extended Nullstellensatz') of Buss *et al.* [18] which can be p-simulated by WF but it is not known if EF p-simulates it too. The classes of proof system that may be stronger than EF are, for example, quantified propositional calculi G_i of Krajíček and Pudlák[74] corresponding to theories T^i_2 or implicit proof systems like iEF of Krajíček [65] corresponding even to $S^1_2 + \text{Exp}$ or stronger theories. In fact, any first-order theory T capable of formalization of logic and having a p-time set of axioms (e.g. PA or ZFC) can be interpreted as a proof system: a proof of tautology τ is any T-proof of the formula in the language of T formalizing that 'τ is a tautology'. But an arbitrary proof system Q can be p-simulated by EF if we add translations $||\text{Ref}_Q||^n$, $n \geq 1$, of the reflection principle for Q (here we use the first-order notation for translations). This set of tautologies is p-time and sparse. The corresponding theory would be an extension of PV by a true $\forall\Pi^b_1$-sentence (see Krajíček and Pudlák [73]).

Third, in the propositional world, EF has a similar role here as a 'sufficiently strong base theory' has in the first-order world. Some constructions (axiomatizations, p-simulations, constructions of short proofs, etc.) require that a proof system is sufficiently strong, and EF functions very smoothly (owing to its correspondence with PV) in this respect. It is true that one can often take for the base system something much weaker (frequently even resolution) but it may make the particular construction in question more cumbersome.

Fourth, EF has also the technical advantage of being p-equivalent to a variety of other interesting proof systems, including Extended Resolution ER, Substitution Frege SF (and other variants of Frege) or a quantified system G^*_1.

All in all, we believe that the choice of EF as a system personifying 'a strong proof system' is an honest one (i.e. no hidden restrictions) but which does not introduce overly abstract notions.

See Krajíček and Pudlák [73] or Krajíček [56] for more detailed expositions of the correspondence between general proof systems and bounded arithmetic theories.

29

Proof complexity generators: definitions and facts

We have chosen the model-theoretic approach (Chapter 18 and Theorem 28.2.1) as a framework in which to think about lower bounds for EF. However, an all-important ingredient is also to have examples of specific candidate tautologies that could be plausibly conjectured to be hard for EF. The experience with lower bound proofs for weaker proof systems shows that one needs specific, combinatorially transparent, formulas to work with. From a model-theoretic perspective this is not surprising: a lower bound is, in effect, a construction of a model where the formula is not valid, so in order to hope to describe such a model we had better understand what the formula means. Of course, one may expect that a random 3DNF (with n variables and $c \cdot n$ clauses, for suitable c) is hard for any proof system, and for some specific systems this has indeed been established (see e.g. Chvatal and Szemeredi [22]). However, lower bounds for random formulas are obtained by verifying that in a lower bound proof discovered originally for some different and specific formula the only properties of the formula that are used are those that are also shared by the random one.

Unfortunately the formulas that worked for weak proof systems are demonstrably easy to prove in EF. The foremost example is the PHP principle we have seen earlier (Chapter 19). However, PHP has short proofs in EF (see Cook and Reckhow [31]) and, in fact, even in F (see Buss [12]). Similar remarks apply to all combinatorial principles considered earlier, in particular to a variety of the so-called counting principles.

A class of tautologies that is provably hard for all non-optimal proof systems is that of tautologies formalizing reflection principles. If P is a non-optimal proof system and Q is a proof system that P does not simulate (see Chapter 27 for these notions) then formulas $||\mathrm{Ref}_Q||^n$ formalizing the soundness of Q with respect to proofs of length at most n are demonstrably hard for P. This goes back to Cook [28] and Krajíček and Pudlák [73] and is exposed in detail in Krajíček

[56]. The drawback of these formulas is that one derives their hardness for P from another lengths-of-proofs assumption, namely that Q cannot be simulated by P.

A very interesting class of tautologies expressing the disjointness of suitable examples of disjoint NP pairs was considered in connection with feasible interpolation; see the discussion in Chapter 27. EF and stronger proof systems do not admit feasible interpolation assuming that RSA is secure (see Krajíček and Pudlák [73]) and, in fact, the argument from Krajíček and Pudlák [73] shows that some of these tautologies do have short proofs in EF and hence, assuming their inseparability by polynomial-time sets, the in-feasibility of interpolation for EF is deduced.

Fortunately there is yet another class of formulas proposed as candidates for hard tautologies. These formulas are defined using specific maps called **proof complexity generators**. The resulting formulas, the so-called **τ-formulas**, could conceivably be hard even in very strong proof systems, including EF. In fact, for all we know at present, some τ-formulas may be hard for all proof systems. The τ-formulas have been defined by Krajíček [61] and independently by Alekhnovich *et al.* [5], and there is now an emerging theory (see papers by Krajíček and by Razborov: [62, 96, 63, 96, 64, 66, 69, 70]); the introductions to Krajíček [63] and to Razborov [96] offer a comprehensive discussion of motivations, aims and background (from slightly different perspectives though).

The rest of the chapter is devoted to an exposition of a part of this theory. Section 29.1 introduces the construction of τ-formulas and the definition of the hardness of a map, a proof complexity generator. Then we discuss three specific maps deemed at present as possibly hard for strong proof systems: the truth-table function (Sections 29.2, 29.3 and 30.1), the Nisan–Wigderson generator (Section 29.4), and the gadget generator (Section 29.5).

29.1 τ-formulas and their hardness

We shall consider maps

$$g : \{0,1\}^n \to \{0,1\}^m$$

computed by a family of polynomial-size circuits $\{C_n\}_n$, and we shall assume that $m = m(n)$ is an injective function of n and that $m = m(n) > n$ (the injectivity of $m(n)$ implies that any string is in the range of at most one circuit C_n and that is convenient). Such maps will be called P/poly, p-stretching.

As the domain $\{0,1\}^n$ of C_n is smaller than $\{0,1\}^m$, the range of C_n, $\mathrm{Rng}(C_n)$, is a proper subset of $\{0,1\}^m$. Hence there are strings $b \in \{0,1\}^m \setminus \mathrm{Rng}(C_n)$ and for any such b we define the **τ-formula** $\tau(C_n)_b(x)$ as:

$$b \not\equiv C_n(x),$$

expressing that $b \notin \mathrm{Rng}(C_n)$ (and hence also expressing that $b \notin \mathrm{Rng}(g)$ as $\mathrm{Rng}(g) \cap \{0,1\}^m = \mathrm{Rng}(C_n)$) in the sense that

$$\tau(C_n)_b \text{ is a tautology iff } b \notin \mathrm{Rng}(C_n).$$

Formula $\tau(C_n)_b(x)$ is a disjunction:

$$\bigvee_{i \leq m} b_i \not\equiv C_{n,i}(x),$$

where b_i is the ith bit of b and $C_{n,i}$ is the (sub)circuit of C_n computing the ith bit of its output. The variables of $\tau(C_n)_b$ are n variables $x_1, \ldots, x_n$ for bits of a potential element x of $\{0,1\}^n$, and also $n^{O(1)}$ many auxiliary variables used in order to define the computation of C_n on x. When circuits C_n are thought of as canonically determined by g we shall denote the formulas simply $\tau(g)_b$.

The main aim is to find maps g for which the resulting τ-formulas are hard to prove in as strong proof systems as possible. To measure the hardness, we introduce, following Razborov [96], the following definition.

Definition 29.1.1 Let g be a P/poly, p-stretching map and let P be a proof system. When only finitely many $\tau(g)_b$ have P-proofs bounded in size by any given polynomial (resp. sub-exponential) bound we say that g is a **hard** (resp. **exponentially hard**) proof complexity generator for P.

Note that the size of the τ-formulas is polynomial in m which is polynomial in n as well because of our stipulation that g is P/poly. However, the length of the formulas would remain polynomial in m even if we allowed m to be super-polynomially bigger than n and the circuits computing g to be polynomial in m rather than in n. In addition, the property '$b \notin \mathrm{Rng}(g)$' can be expressed by a tautology even for maps g whose output bits are defined by non-uniform NP$\cap$coNP conditions (in terms of the input bits). In fact, we could again allow the non-deterministic time to be polynomial in m rather than in n.

We now observe that the hardness can be interpreted as a sort of a hitting set property (following Krajíček [63]).

Definition 29.1.2 Let g be a P/poly p-stretching map computed by circuits $\{C_n\}_n$, and let P be a proof system. The **resultant** of g with respect to P, denoted

Res_g^P, is the class of all NP/poly-sets $A \subseteq \{0,1\}^*$ such that for some definition of A

$$y \in A \text{ iff } \exists z(|z| \leq |y|^k) B(y,z)$$

where $B(y,z)$ is a P/poly relation, and the proof system P admits polynomial-size proofs of the propositional statements

$$||\forall x, z\ (B(y,z) \rightarrow C_n(x) \neq y)||^{m(n)}$$

(with the bounds to the lengths of x and z implicitly polynomial in n).

Lemma 29.1.3 (Krajíček [63]) *Let g be a* P/poly p-*stretching map computed by circuits $\{C_n\}_n$. Let* P $\supseteq$ EF *be a proof system. Then the following two conditions are equivalent:*

1. *There exists $t \geq 0$ such that for infinitely many $n \geq 1$ and $b \in \{0,1\}^{m(n)}$ the formula $\tau(C_n)_b$ has a* P*-proof of size $\leq |\tau(C_n)_b|^t = |b|^{O(t)}$.*
2. *The resultant* Res_g^P *contains an infinite set.*

Note also that a P/poly p-stretching map g whose range would intersect all infinite NP-sets would be a proof complexity generator hard for any proof system P. This is because for any P and $k \geq 1$ the set of strings b for which the corresponding τ-formula has a P-proof of length at most m^k is in NP.

Similarly, if $\mathrm{Rng}(g)$ intersected all NP-sets of density at least $\frac{1}{2}$ it would still follow that for any P and any $k \geq 1$ there is at least one $b \notin \mathrm{Rng}(g)$ for large enough m for which $\tau(g)_b$ cannot be proved by a P-proof of size less than m^k. This is because the ambient space $\{0,1\}^m$ is at least twice as large as $\mathrm{Rng}(g) \cap \{0,1\}^m$.

In the next few sections we shall introduce three candidates for hard proof complexity generators. Let us remark that it is not true that any potential strong pseudo random number generator g (SPRNG) will make a good candidate to be a hard proof complexity generator (by a SPRNG we mean a function obeying Definition 25.6 but with respect to non-uniform adversaries; see a remark before Definition 25.6). For example, assume f is a one-way permutation and g is a SPRNG constructed from f by adding its hard bit. Assuming that EF gives a short proof that f is injective it is easy to see that it also gives short proofs of all τ-formulas from g. This applies, for example, to RSA as shown in Krajíček and Pudlák [73].

However, there is some use for SPRNGs in this context nevertheless. As Alekhnovich *et al.* [5] noted, if $n < m/2$ and g is a SPRNG, then map

$$h : (x^1, x^2) \in \{0,1\}^n \times \{0,1\}^n \rightarrow \{0,1\}^m$$

defined by the bit-wise sum

$$h(x^1, x^2) := g(x^1) \oplus g(x^2)$$

will be a hard proof complexity generator for any proof system that admits feasible interpolation. The proof of this fact is by contradiction: the existence of a short proof of some τ-formula $\tau(h)_b$ together with feasible interpolation would allow to define a P/poly set C^* disjoint with Rng(g) but of measure at least $\frac{1}{2}$, contradicting the pseudo randomness of g. The construction of C^* is quite analogous to the argument in the construction of C^* in the proof of Theorem 29.2.3 below.

29.2 The truth-table function

The truth-table function is the most prominent example of a proof complexity generator.

Definition 29.2.1 Let $k \geq 1$ be a parameter and let $s = s(k) \leq 2^{(1-\Omega(1))k}$ be a function of k.

The **truth-table function $\mathbf{tt}_{s,k}$** takes as an input a description of a circuit C with k inputs and of size at most s, and outputs the 2^k bits of the truth table of the Boolean function that C computes.

A size s circuit is encoded by $O(s \log s) < 2^k$ bits. Hence $\mathbf{tt}_{s,k}$ maps $n := O(s \log s)$ bit strings to $m := 2^k > n$ bit strings.

$\mathbf{tt}_{s,k}$ is, by definition, equal to the string of zeros at inputs that do not encode a size $\leq s$ circuit with k inputs.

A basic observation is that the τ-formula $\tau(\mathbf{tt}_{s,k})_b$ expresses that $b \in \{0,1\}^{2^k}$ is the truth table of a Boolean function which cannot be computed by a circuit of size at most s. That is, the τ-formula expresses a circuit lower bound for b.

This interpretation of the formulas allows us to make two observations.

Theorem 29.2.2 *(1) Assume there is a proof system* P *and* $s(k) \geq 2^{\Omega(k)}$ *such that* $\mathbf{tt}_{s,k}$ *is not hard for* P. *Then* BPP $\subseteq$ NP.

(2) Assume there is a proof system P *and* $s(k) \geq k^{\omega(1)}$ *such that* $\mathbf{tt}_{s,k}$ *is not hard for* P. *Then* NEXP $\not\subseteq$ P/poly.

Proof: For (1) consider[1] the following process: guess a pair (b, π), where b is the truth table of a function outside of the range of $\mathbf{tt}_{s,k}$ (i.e. with exponential

[1] See Krajíček [63]; the observation is attributed to R. Impagliazzo there.

circuit complexity) and π is a polynomial size (i.e. $2^{O(k)}$) P-proof of $\tau(\mathbf{tt}_{s,k})_b$. By Nisan and Wigderson [84] and Impagliazzo and Wigderson [48] such a function b can be used for derandomization of BPP.

Statement (2) follows, as Impagliazzo, Kabanets and Wigderson [44] proved (by derandomizing MA) that the existence of an NP-way (considered there as a form of NP-natural proofs) of certifying a superpolynomial circuit complexity of a function implies that NEXP $\not\subseteq$ P/poly. The hypothesis of (2) provides such a way of certification. □

The statement can be interpreted as saying that it is not to be expected that it will be simple to exhibit a proof system for which the truth-table function is not a hard proof complexity generator. Note also that, while it is easy to prove by a counting argument that there exists a function b of a superpolynomial circuit complexity, i.e. $b \notin \mathrm{Rng}(\mathbf{tt}_{s,k})$ for $s \geq k^{\omega(1)}$, we do not know how to give a short proof of this (in polynomial size) for any particular b in the whole of Mathematics (as formalized e.g. by ZFC).

We shall now make an observation in the opposite direction; we shall describe a fairly large class of proof systems for which the truth-table function is hard. By a strong pseudo random generator (SPRNG) in the next statement we mean the same thing as at the end of Section 29.1. The proof of the theorem uses a trick invented by Razborov [93].

Theorem 29.2.3 *Assume that there exists a* SPRNG. *Let* $s(k) \geq k^{\omega(1)}$. *Then the truth-table function* $\mathbf{tt}_{s,k}$ *is hard for any proof system that admits feasible interpolation.*

Proof: Assume $\mathbf{tt}_{s,k}$ is not hard for P and let $b \in \{0,1\}^{2^k}$ be a function such that P admits a polynomial-size proof of

$$\tau(\mathbf{tt}_{s,k})_b.$$

Put $t := s/3$ and let x be a 2^k-tuple of atoms. Then P also proves in polynomial size the disjunction

$$\tau(\mathbf{tt}_{t,k})_x \ \vee\ \tau(\mathbf{tt}_{t,k})_{b\oplus x}$$

as two circuits of size $\leq t$ computing some function x and $b \oplus x$ respectively can be combined into a circuit of size $\leq s$ computing b.

By feasible interpolation for P there is a circuit $C(y)$ with 2^k inputs and of size $2^{O(k)}$ defining a subset $C^* \subseteq \{0,1\}^{2^k}$ that separates two sets

$$A := \{x \in \{0,1\}^{2^k} \mid x \in \mathrm{Rng}(\mathbf{tt}_{t,k})\}$$

and

$$B := \{x \in \{0,1\}^{2^k} \mid x \oplus b \in \mathrm{Rng}(\mathbf{tt}_{t,k})\}$$

in the sense that

$$A \subseteq C^* \text{ and } C^* \cap B = \emptyset.$$

If C^* contains at most half of $\{0,1\}^{2^k}$ put $D(y) := \neg C(y)$ otherwise put $D(y) := C(y \oplus b)$. It follows that if y satisfies D then it is a truth table of a function that cannot be computed by a circuit of size $\leq t$. In particular, it cannot be computed by a polynomial-size circuit.

Such a circuit D is a P/poly-natural property against P/poly in the sense of Razborov–Rudich's natural proofs [97]. But by the main theorem of [97] the existence of such a property implies that no SPRNG exists. That is a contradiction. □

The hardness of $\mathbf{tt}_{s,k}$ is also known to follow for a proof system if a version of the weak pigeonhole principle with $2^{n^{\Omega(1)}}$ pigeons and n holes is hard for the proof system; see Razborov [93]. But this condition does not apply even to constant-depth Frege systems.

29.3 Iterability and the completeness of $\mathbf{tt}_{s,k}$

For the next definition and for Theorem 29.3.2 we will assume that h is a P/poly map that stretches inputs of length n to outputs of length $n+1$ (that is the most general context for Definition 29.3.1 and Theorem 29.3.2). We shall denote by h_n the restriction of h to $\{0,1\}^n$.

Definition 29.3.1 (Krajíček[63]) Let $s(n) \geq 1$ be a function. A function h is $s(n)$-**iterable** for proof system P iff all disjunctions of the form

$$\tau(h_n)_{b_1}(x^1) \vee \cdots \vee \tau(h_n)_{b_t}(x^t)$$

require P-proofs of size at least $s(n)$. Here $t \geq 1$ is arbitrary, and $b_1, \ldots, b_t$ are $(n+1)$-tuples of variables and constants such that

1. x^i are disjoint n-tuples of atoms, for $i \leq t$,
2. $b_1 \in \{0,1\}^{n+1}$ (i.e. no variables in b_1),
3. variables occurring in b_i are among $x^1, \ldots, x^{i-1}$, for $i \leq t$.

If map h satisfies a variant of the definition where b_i are allowed to be any circuits (with $n+1$ outputs) obeying the restrictions 1–3 it is called $s(n)$-**pseudo-surjective** for P.

Note that the $s(n)$-pseudosurjectivity with super-polynomial (resp. exponential) $s(n)$ implies the (exponential) iterability which in turn implies the (exponential) hardness, the latter being the iterability condition with $t = 1$.

The disjunction in the definition can be interpreted as follows (see Krajíček [63, Section 3] for a more extensive commentary including an interpretation using an interactive computation between Student and Teacher in Krajíček, Pudlák and Sgall [77]). Assume that the disjunction is a tautology. Then it may be that already the first disjunct $\tau(h_n)_{b_1}(x^1)$ is a tautology, meaning that the string b_1 is outside of the range of h_n. If not, and $a^1 \in \{0,1\}^n$ is such that $h_n(a^1) = b_1$, then $b_2(a^1)$ is the next candidate for a string being outside of the range of h_n, etc. Because the disjunction is a tautology we must find in this process a string outside of the range of h_n in at most t rounds.

The importance of the concept stems from the next theorem.

Theorem 29.3.2 (Krajíček [63]) *Assume that proof system* P *simulates resolution. Then the following two statements hold:*

1. *There exists an h (exponentially) iterable for* P *iff, for any* $0 < \delta < 1$, *the truth-table function* $\mathbf{tt}_{s,k}$ *with* $s = 2^{\delta k}$ *is (exponentially) iterable for* P *too.*
2. *There exists an h that is exponentially iterable for* P *iff there is* $c \geq 1$ *such that for* $s = k^c$ *the truth-table function* $\mathbf{tt}_{s,k}$ *is exponentially iterable for* P.

The same holds with iterability replaced by pseudosurjectivity.

In other words, if we replace hardness by iterability then the truth-table function is the 'hardest' function.

29.4 The Nisan–Wigderson generator

We pointed out at the end of Section 29.1 that not all maps that are considered as plausible candidates for strong pseudorandom generators (SPRNGs) are also good candidates for hard proof complexity generators. However, the argument outlined there for why some SPRNGs are not hard from the proof complexity point of view does not seem to apply to the classic Nisan–Wigderson generator (NW generator for short); see Nisan and Wigderson [84]. In fact, the NW generator was proposed as a possibly hard proof complexity generator in Alekhnovich *et al.* [5] and taken up in Krajíček[63], although the motivations and, more importantly, the choice of parameters in the two proposals were different.

An NW generator is defined by a 0–1 matrix A and a Boolean function f. In computational complexity theory the term **NW generator** usually refers to the original construction of Nisan and Wigderson [84], i.e. where A is the so-called (d, ℓ) design (see below) and f is a suitably hard function. Here we shall use the term more loosely, meaning any map defined in the following manner.

Let $n < m$ and let A be an $m \times n$ 0–1 matrix with ℓ 1s per row. Then $J_i(A) := \{j \leq [n] | A_{ij} = 1\}$. Let $f : \{0,1\}^\ell \to \{0,1\}$ be a Boolean function. Define function $\mathrm{NW}_{A,f} : \{0,1\}^n \to \{0,1\}^m$ as follows: the ith bit of the output is computed by f from the bits of the input that belong to $J_i(A)$. Matrix A is an **(d, ℓ)-design** if in addition the intersection of any two different rows $J_i(A) \cap J_k(A)$ has size at most d.

Note that the size of the τ-formula from $NW_{A,f}$ is $m^{O(1)}$ if f is computed by a circuit of size $m^{O(1)}$. In fact, that remains true even if f is computed in the non-uniform class $\mathrm{NTime}(m^{O(1)}) \cap \mathrm{coNTime}(m^{O(1)})$.

The idea that $\mathrm{NW}_{A,f}$ forms a good proof complexity generator has been formulated in Alekhnovih *et al.* [5] in general terms, assuming that A has a suitable combinatorial property and that f is hard in some sense for the proof system in question. The combinatorial property used in Alekhnovich *et al.* [5] was formulated as an expansion requirement, using the following definition.

Definition 29.4.1 (Alekhnovich *et al.* [5, Definition 2.1]) Let A be an $m \times n$ matrix. A boundary of a set of rows $I \subseteq [m]$, denoted $\partial_A(I)$, is the set of $j \in [n]$ such that entry A_{ij} equals 1 for exactly one $i \in I$.

Let $1 \leq r \leq m$, $\ell \leq n$ and $c > 0$ be any parameters. Matrix A is an (r, ℓ, c)-**expander** iff A has ℓ 1s per row and, for all $I \subseteq [m]$, $|I| \leq r$, $|\partial_A(I)| \geq c|I|$.

For various arguments in Alekhnovich *et al.* [5] and in Krajíček [63] one needs r close to n and $c = \Omega(\ell)$. As shown in Alekhnovich *et al.* [5], a random ℓ-sparse matrix has the expansion property with suitable parameters but not an arbitrary $(\Omega(\ell), \ell)$-design.

The hardness of $\mathrm{NW}_{A,f}$ was shown in Alekhnovich *et al.* [5] for expanding matrices A with $m \leq n^{2-\Omega(1)}$ and suitable f (e.g. parity function) for resolution (and with a different definition of the τ-formulas also for polynomial calculus and its combination with resolution PCR). This was then improved to $m = n^{3-\Omega(1)}$ and $\ell = O(1)$ in Krajíček [63], and subsequently Razborov [96] has shown that such maps $\mathrm{NW}_{A,f}$ are exponentially iterable for resolution (and for its extension $R(k)$ with small k). In fact, Razborov [96] also observed that the argument in Krajíček [63] already shows the exponential iterability too.

29.5 Gadget generators

Start with a polynomial-time function

$$f \,:\, \{0,1\}^{\ell} \times \{0,1\}^{k} \to \{0,1\}^{k+1}$$

and define a new function

$$g \,:\, \{0,1\}^{n} \to \{0,1\}^{n+1}$$

for

$$n \,:=\, \ell + k \cdot (\ell + 1)$$

as follows:

- An input $x \in \{0,1\}^n$ is interpreted as one string v of length ℓ and followed $\ell + 1$ strings u^s of length k, $s = 1, \ldots, \ell + 1$.
- The output $y \in \{0,1\}^{n+1}$ is the concatenation of $\ell + 1$ strings y^s, each of length $k + 1$, defined by

$$y^s := f(v, u^s).$$

Each y^s has one extra bit of length over that of u^s, so the first ℓ of the blocks swallow v and the last one produces the one extra bit of output we want. The string v is called **a gadget**.

Of course, for this to work we have to have a suitable gadget and a function f. We now present a simple example constructed in Krajíček [70]. The gadget there is interpreted as a graph of a function h between $[k]$ and $[k+1]$ and the idea is that unless we can prove PHP in a proof system P we cannot rule out in P that h is a bijection and one may use it to stretch each block u^s to a block y^s longer by one bit.

Definition 29.5.1 Let $k, t \geq 1$ be any parameters such that $t > k \cdot (k+1)$. Put $n := k \cdot (k + 1 + t)$ and $m := (k+1) \cdot t$. Hence $m > n$.

Map $g_{k,t} : \{0,1\}^n \to \{0,1\}^m$ is defined as follows. Interpret input string x of length n as

$$x = (v, u^1, \ldots, u^t)$$

where

$$v = (v_{ij})_{i \in [k+1], j \in [k]} \text{ and } u^s = (u^s_j)_{j \in [k]}$$

for $s = 1, \ldots, t$.

The output string y of length m is defined as $y := (y^1, \ldots, y^t)$ where

$$y_i^s := \bigvee_{j \in [k]} (v_{ij} \wedge u_j^s)$$

for $s = 1, \ldots, t$.

Note that the map f in this particular construction is very simply defined: its output bits are defined by 2DNF formulas.

Theorem 29.5.2 (Krajíček [70]) *Let $d \geq 2$ and assume $k \geq 1$ and $t = k^2 + k + 1$. Then, with $n := k \cdot (k + 1 + t)$ as above, the map*

$$g_{k,t} : \{0,1\}^n \to \{0,1\}^{n+1}$$

is an exponentially hard proof complexity generator for constant-depth Frege systems F_d.

Proof: Let $b := (b^1, \ldots, b^t) \in \{0,1\}^{n+1}$ be an arbitrary string, where bs are blocks of length $k+1$, and assume that we have an F_d-proof π of the τ-formula $\tau(g)_b$. Substitute in π everywhere for all variables u_j^s-formulas

$$u_j^s(v) := \bigvee_{i \in [k+1]} v_{ij} \wedge b_i^s.$$

Denote the substituted proof π^v.

Let $\neg\mathrm{PHP}_k(v)$ be the formula expressing that v defines a graph of a function violating the pigeonhole principle from $[k+1]$ into $[k]$ (not necessarily bijective):

$$\bigwedge_{i \in [k+1]} \bigvee_{j \in [k]} v_{ij} \wedge \bigwedge_{i \in [k+1]} \bigwedge_{j_1 \neq j_2 \in [k]} \neg v_{ij_1} \vee \neg v_{ij_2},$$

$$\wedge \bigwedge_{i_1 \neq i_2 \in [k+1]} \bigwedge_{j \in [k]} (\neg p_{i_1 j} \vee \neg p_{i_2 j}).$$

It is not difficult to see that there is a size $n^{O(1)} = k^{O(1)}$ F_d-proof σ of

$$g_{k,t}(v, u^1(v), \ldots, u^t(v)) \neq b \;\to\; \mathrm{PHP}_k(v).$$

If we combine proofs π^v and σ we get a proof of $\mathrm{PHP}_k(v)$. By Ajtai [2], Krajíček *et al.* [79] and Pitassi *et al.* [88] any such proof must have size exponential in k, and hence π^v, and π, must be exponential too. □

Quite generally, one can fix the function f to be the circuit-value function

$$\mathrm{CV}_{\ell,k}(v,u)$$

that takes ℓ bits v describing a circuit C with k input bits and $k+1$ output bits and $u \in \{0,1\}^k$, and outputs the string $C(u)$. It may seem at first that the value of the parameter ℓ does not allow for a canonical choice. However, using a form of self-reducibility of the resulting generators Krajíček [70] observed that it is possible to assume without loss of generality that $\ell \leq k^2$ and, in fact, even $\ell \leq k^{1+\epsilon}$, for any fixed $\epsilon > 0$.

Using as the gadget a map violating PHP results in a generator that is hard for any proof system where PHP is hard (see Krajíček [70] for a discussion of algebraic proof systems). One may use other well-known tautologies that are hard for some proof systems to create a combinatorial situation that can be used as a good gadget. In Section 30.3 we will use as gadgets the data A, f defining an NW-generator but here we give just one more combinatorial example to illustrate the definition (but provably not useful for EF, however).

Assume that $v : [k^2] \to [k^3]$, $w : [k^3] \to [k]$ are two maps defined by their graphs (i.e. we have k^5 and k^4 variables v and w, respectively), and let $G = ([k^3], E)$ be an undirected graph without loops, with vertex set $[n]$ and edges E ($(k^3)(k^3-1)/2$ variables). The Clique/Coloring principle says that it cannot happen simultaneously that:

1. v is an injective map,
2. the range of v is a clique in G,
3. w is a coloring of G.

This is obviously true as otherwise we could color a clique of size k^2 by k colors, which is impossible.

Assume, however, that v, w and E do violate the principle. Then the composed map $w \circ v : [k^2] \to [k]$ is injective and hence violates the weak PHP. In other words, if the weak PHP has only long P-proofs then the generator resulting in the described way from this gadget will be hard for P. An example of a proof system for which PHP is easy while the Clique/Coloring principle is hard is the cutting planes proof system CP; see Pudlák [89].

29.6 Optimal automatizer for the τ-formulas

Hirsch and Itsykson [42] introduced a concept of heuristic randomized proof systems that do make errors, but only a few, and a corresponding concept of an automatizer. We shall recall only the latter as it is simpler and quite natural in the context of proof complexity generators.

Let $L \subseteq \{0,1\}^*$ be a language and $D = (D_n)_{n \geq 1}$ a family of probabilistic distributions on $\{0,1\}^n \setminus L$. Such a pair is called a **distributional proving problem** in Hirsch and Itsykson [42]. Informally, an automatizer for (D, L) is a probabilistic algorithm that accepts all of L and only a little of its complement, when the qualification 'little' is defined via the family D.

Definition 29.6.1 (Hirsch and Itsykson [42]) An automatizer A for a distributional proving problem (D, L) is a probabilistic algorithm that:

1. has as inputs $(x, d) \in \{0,1\}^* \times \mathbf{N}$,
2. on every input either halts and outputs 1 or does not halt at all,
3. for every $(x, d) \in L \times \mathbf{N}$, $A(x, d) = 1$ with probability 1,
4. for any $n, d \geq 1$:

$$\text{Prob}_{x \in D_n} \, [(\text{Prob}[A(x,d) = 1] > 1/4)] \; < \; 1/d$$

where the inner probability is computed over the random choices of A and in the outer probability the x of length n is chosen according to D_n.

The time an automatizer takes on an input is defined to be the median time (the minimal time such that there is at least a fifty – fifty chance that the algorithm halts). An automatizer is **optimal** if it is at most polynomially slower than any other automatizer.

Theorem 29.6.2 (Hirsch and Itsykson [42]) *Let L be a recursively enumerable language and D a family of polynomial-time samplable probability distributions on its complement.*

Then there is an optimal automatizer for the distributional proving problem (D, L).

In general, for an r.e. language, there may be a variety of distributions on the complement of L, and there does not seem to be a natural and universal criterion on how to select one. But in the case of proof complexity generators there does appear to be a most natural choice.

Namely, let g be a polynomial-time function mapping $\{0,1\}^n$ into $\{0,1\}^m$, where $n \to m = m(n)$ is an injective function (and hence m determines n) as in Section 29.1. Define language L^g (obviously r.e.) by:

$$L^g \; := \; \{0,1\}^* \setminus \text{Rng}(g).$$

Now note that the function g itself provides us a with a (p-time samplable) distribution D^g on $\{0,1\}^* \setminus L^g$, i.e. on Rng(g). That is, the distribution D^g_m on $\{0,1\}^m \setminus L^g = \{0,1\}^m \cap \text{Rng}(g)$ gives to $w \in \{0,1\}^m \cap \text{Rng}(g)$ the probability:

$$\text{Prob}_{x \in \{0,1\}^n}[g(x) = w].$$

Theorem 29.6.2 in this context provides an optimal automatizer for accepting the valid $\tau(g)$-formulas. Each formula $\tau(g)_b$ is represented by b and hence L^g represents the set of valid $\tau(g)$-formulas.

Theorem 29.6.3 *Let g be a polynomial-time function mapping $\{0,1\}^n$ into $\{0,1\}^m$, where $n \to m = m(n)$ is an injective function. Then there is an automatizer A that accepts all of $L^g \times \mathbf{N}$ with probability* 1, *for which*

$$\text{Prob}_{x \in \{0,1\}^n} \; [(\text{Prob}[A(g(x),d) = 1] > 1/4)] \; < \; 1/d$$

and which is optimal among all automatizers for the distributional proving problem (D^g, L^g) determined by g.

30

Proof complexity generators: conjectures

All three generators that we have discussed in Chapter 29 are determined by some kind of data that can vary: an NW generator is determined by a matrix A and a function f, a gadget generator by its gadget, and the truth-table function $\mathbf{tt}_{s,k}$ by its size function $s = s(k)$.

A generator with specific data may be hard for one proof system and easy for another one. For example, if f is the parity function then EF can prove all valid τ-formulas for the NW generator by following Gaussian elimination. Or the PHP-gadget is hard for F_d while easy for CP. With the current sorry state of circuit complexity lower bounds one has to take s very slow indeed to have a similar example (but $s(k) = 2k$ will do).

It is an interesting question whether one can choose these data in some particular way so that the resulting generator would be hard for *all* proof systems. In this chapter we present various informal speculations as well as formal conjectures related to the hardness of these generators.

Theorem 29.3.2 makes in a sense the truth-table function the most prominent among proof complexity generators. We will thus start our speculations in the first section with this function.

30.1 On provability of circuit lower bounds

Assume that a lower bound $s = s(k)$ (maybe even with an exponential s) is valid for SAT. Then the truth-table function $\mathbf{tt}_{s,k}$ will not be hard for all proof systems. Define a proof system P which, upon receiving a string b of length 2^k, checks whether b is or is not equal to χ_{SAT_k}, the characteristic function of SAT restricted to inputs of length k. If so, then it accepts $\tau(\mathbf{tt}_{s,k})_b$ as a tautology, and if not it proceeds as, say, EF. The point is that deciding the required property of

b can be done in polynomial time and hence P is indeed a proof system in the sense of Cook and Reckhow.

Same argument applies to all languages L in place of SAT that belong to the exponential class E. In fact, if L is in NE $\cap$ coNE one can still define in an analogous way a proof system Q for which $\mathbf{tt}_{s,k}$ will not be hard: this time Q expects that all non-deterministic witnesses for all values in the truth table b are part of a proof it is given. Each individual witness has size a polynomial in 2^k and there are 2^k of them, so the collection of all required witnesses is of polynomial size (and can be verified as correct in polynomial time).

One may speculate that this is the *only* way that $\mathbf{tt}_{s,k}$ may fail to be hard for a proof system.

Possibility A: *Assume that* $\mathbf{tt}_{s,k}$ *is not hard for a proof system* P. *Then there is* $L \in$ NE $\cap$ coNE *such that* P *admits polynomial-size proofs of tautologies*

$$\tau(\mathbf{tt}_{s,k})_{\chi_{L_k}}.$$

An obvious consequence is the following observation. For a function $s = s(k)$ denote by Size(s) the class of languages with circuit complexity at most $s(k)$.

Lemma 30.1.1 *Assume that Possibility A is true, and that* $s(k) \leq 2^{(1-\Omega(1))k}$. *Then* NE $\cap$ coNE $\subseteq$ Size(s) *implies* NP $\neq$ coNP.

Note, as a small test of Possibility A, that the implication

$$\text{NE} \cap \text{coNE} \subseteq \text{Size}(s) \;\rightarrow\; \text{NP} \neq \text{coNP}$$

is actually true, as NP = coNP implies that NE $\cap$ coNE contains a language of any circuit complexity $2^{(1-\Omega(1))k}$ (as in the proof of Theorem 30.2.2).

Another speculation concerns a possibility to give a sufficient condition on a proof system guaranteeing that $\mathbf{tt}_{s,k}$ is hard for it.

Possibility B: *If a proof system* P *is not optimal then for all* $\delta > 0$ *the function* $\mathbf{tt}_{s,k}$ *with* $s(k) = 2^{\delta \cdot k}$ *is hard for* P.

The non-optimality of P implies (via known links between simulations and reflection principles) that P cannot refute that another proof system Q is not sound, say proves 0. Without loss of generality Q is EF augmented by a polynomial-time set of formulas as extra axioms (see Krajíček [56]). Now, a proof of 0 in, for example, EF itself is essentially the same thing as a description of a circuit without inputs and then two different 'deductions' of its value leading to opposite values. For EF the meaning of 'a deduction' is just reasoning in resolution R, for Q stronger than EF it is R enhanced in a particular way (depending on Q). The idea behind Possibility B is that such a 'contradictory' circuit might be transformed into a circuit (perhaps with an almost exponential blow-up) computing any predetermined function b. In particular, P cannot give a short proof of $\tau(\mathbf{tt}_{s,k})_b$.

A small part of Possibility B follows from Krajíček [63, Theorem 5.2]: it is shown there that if EF does not simulate WF (a proof system possibly stronger that EF, introduced by Jeřábek [50]) then $\mathbf{tt}_{s,k}$ is not only hard but is even pseudo-surjective for EF.

Lemma 30.1.2 *Assume that Possibility B is true and that* $E \not\subseteq \text{Size}(2^{o(n)})$. *Then there exists an optimal proof system.*

Proof: Let L be a complete language for E and $s(k)$ be some specific exponential lower bound $2^{\Omega(k)}$ to its circuit complexity (by the hypothesis of the lemma). Form a proof system P that is EF augmented by all instances of τ-formulas

$$\tau(\mathbf{tt}_{s,k})_{L_k}.$$

□

We now turn to a particular problem that seems to have nothing to do with proof complexity. Consider set CS $\subseteq \{1\}^* \times \{0,1\}^*$ (for **Circuit Size**) consisting of all pairs

$$(1^{(s)}, b),$$

where b is a string of length 2^k defining a Boolean function in k variables of circuit complexity at most s, and $s \leq 2^k$.

Problem 30.1.3 *Is set* CS NP-*complete?*

This problem has been studied in a slightly different version by Kabanets and Cai [51], under the name 'circuit minimization problem'. The difference is that they do encode s in binary. That is, of course, irrelevant from the point of view of mutual p-reducibility. But Kabanets and Cai [51] consider a notion of 'natural reduction' (i.e. length determined) and then the unary and the binary encodings seem to matter.

Possibility C: CS *is* NP-*complete and moreover there are* p-*time functions* f, g *and* h *such that:*

1. f *is defined on* 3CNF *formulas* φ *and it is a* p-*reduction of* 3SAT *to* CS.

2. *g is defined on pairs (φ, a) of a* 3CNF *formula and a truth assignment, its value is a circuit, and it holds that*

$$\varphi(a) = 1 \rightarrow \mathbf{tt}_{s,k}(g(\varphi, a)) = \pi_2(f(\varphi)),$$

where π_2 is the projection on the second coordinate ($f(\varphi)$ is a pair $(1^{(s)}, b)$).

3. *h is defined on pairs (C, b) of a circuit and a truth table, its value is a truth assignment, and the following holds. If f maps a* 3CNF *formula φ to pair $(1^{(s)}, b)$ then:*

$$\mathbf{tt}_{s,k}(C) = b \rightarrow \varphi(h(C, b)) = 1.$$

In other words, maps g and h send witnesses to witnesses of the two NP*-properties.*

Define a proof system P as the extension of EF by all propositional translations (for all lengths $n \geq 1$) of statements 2 and 3 from Possibility C. Owing to the strength of PV it seems likely that, if possibility C holds, statements 2 and 3 may be provable in PV. Hence, owing to the relation of EF to PV, these extra axioms would be polynomially provable in EF and hence, in fact, P = EF would hold.

In the proof of the following lemmas we use known properties of the propositional $|| \dots ||$-translation; see Krajíček [56].

Lemma 30.1.4 *Assume Possibility C. Then there is a constant $c \geq 1$ such that every tautology φ of length n has a* P*-proof of size at most n^c from instances of some valid τ-formula $\tau(\mathbf{tt}_{s,k})_b$.*

Informally: circuit lower bounds are the hardest tautologies over a fixed proof system P.

Proof: Let ψ be a tautology and assume without loss of generality that it is a 3DNF formula. Take φ to be the 3CNF of $\neg\psi$ assume $|\varphi| \leq n$ and that atoms of φ are among $x_1, \dots, x_n$.

An instance of one of the new axioms in P is

$$|| \, \varphi(x_1, \dots, x_n) \rightarrow \mathbf{tt}_{s,k}(g(\varphi, x)) = b \, ||^n,$$

where s, k and b are given by $\pi_2(f(\varphi)) \in \{0,1\}^{2^k}$ and $x = (x_1, \dots, x_n.)$

A substitution instance of the axiom $\tau(\mathbf{tt}_{s,k})_b$ is $\mathbf{tt}_{s,k}(g(\varphi, x)) \neq b$, and hence $\neg\varphi$ follows. □

Lemma 30.1.5 *Assume Possibility C. Assume also that $\{\varphi_n\}_n$ is a sequence of tautologies constructed by a polynomial-time algorithm from $1^{(n)}$.*

Then there is a language L in class E and $s = s(k)$ such that τ-formulas $\tau(\mathbf{tt}_{s,k})_{L_k}$ are valid and each φ_n has a polynomial-size P*-proof from instances of one of these τ-formulas.*

Proof: Composing the polynomial-time construction of φ_n from $1^{(n)}$ with the map $\pi_2 \circ f$ gives a polynomial-time algorithm computing the truth table b in the previous lemma from $1^{(n)}$, defining a language L. Hence $L \in$ E. □

Lemma 30.1.6 *Assume Possibility C. Then any propositional proof system Q can be simulated by a proof system* P *augmented by instances of valid τ-formulas $\tau(\mathbf{tt}_{s,k})_{L_k}$, for some L in class E and some functions s.*

Proof: The statement follows from Lemma 30.1.5 using the fact that Q can be p-simulated by EF augmented by the propositional translations of the reflection principle for Q; these formulas are p-time construable. □

Analogously to Lemma 30.1.5 the following statement holds.

Lemma 30.1.7 *Let $\delta > 0$ and take a subset* CS′ *of* CS *consisting of those pairs $(1^{(s)}, b)$ such that $s \geq 2^{\delta k}$. Assume that Possibility C holds for* CS′.

Then there exists a language L in E requiring circuits of size $2^{\Omega(k)}$. (And hence P = BPP*).*

30.2 Razborov's conjecture on the NW generator and EF

Lower bound arguments for the hardness of the NW generator in resolution in Krajíček[63] and Razborov [96], as well as related results in the earlier Alekhnovich *et al.* [5], use a strong expanding property of matrix A that is not guaranteed by being a mere (d, ℓ)-design, with parameters ℓ and d satisfying the constraints of the original construction in Nisan and Wigderson [84]. In fact, only recently Pich [87] demonstrated the hardness of the original NW generator from Nisan and Wigderson [84] for proof systems admitting feasible interpolation (which includes resolution). Nevertheless, Razborov [96] has made the following intriguing conjecture.

Conjecture 30.2.1 (Razborov [96, Conjecture 2]) *Any* NW*-generator based on a matrix A which is a combinatorial design with the same parameters as in Nisan and Wigderson [84] and on any function f in* NP∩coNP *that is hard on average for* P/poly *is hard for* EF.

The parameters are not explicitly specified in Razborov [96] and the formulation 'with the same parameters' requires some investigation into what the constraints actually are.

The conjecture is not valid for all parameters for which the Nisan and Wigderson [84] construction works (see Krajíček [64] for an example). But it seems that the conjecture alludes to the following constraints on the parameters in the main construction of combinatorial designs in Nisan and Wigderson [84, Lemma 2.5]:

$$d = \log(m), \quad \log(m) \leq \ell \leq m, \quad n = O(\ell^2).$$

We shall interpret the phrase 'hard on average for P/poly' from the conjecture as meaning that the hardness on average of f in the sense of Nisan and Wigderson [84] is exponential:

$$(1) \qquad H_f(\ell) \geq 2^{\Omega(\ell)}.$$

The quantity $H_f(\ell)$, **the hardness** of function f, is defined to be the minimal s such that there is a circuit C of size s such that

$$\text{Prob}_u[f(u) = C(u)] \geq 1/2 + 1/s,$$

where $u \in \{0,1\}^\ell$. In addition, Nisan and Wigderson [84, Lemma 2.4] requires that

$$H_f(\ell) \geq m^2.$$

Putting this together and taking the maximal allowed value for m we can write the constrains as:

$$(2) \qquad m = 2^{\epsilon \cdot n^{1/2}}, \quad d = \epsilon \cdot n^{1/2}, \quad \ell = n^{1/2},$$

where $\epsilon > 0$ is a constant. But note that it would be valid to take also the minimal value $m := n+1$ (as we shall do later in Chapter 31).

The conjecture requires that $f \in \text{NP} \cap \text{coNP}$. But note that if we relax this condition to:

$$(3) \qquad f \in \text{NTime}(m^{O(1)}) \cap \text{coNTime}(m^{O(1)}).$$

the size of the τ-formula will remain $m^{O(1)}$ (by the discussion after Definition 29.1.1).

Consider (following Krajíček[64]) instead of the conjecture the following **Statement (R)**, which states the conjecture for the specific parameters (1) and (2) above but poses on the function f only requirement (3).

(R) *Let g be an* NW-*generator based on an $m \times n$ matrix A that is an (d, ℓ) combinatorial design and on any function f such that the constrains in (1), (2) and (3) are satisfied. Then g is hard for* EF.

Statement (R) asserts a conditional lower bound for EF. However, Krajíček [64] made a simple but slightly surprising observation that it, in fact, implies also an unconditional lower bound for EF. We sketch its proof from Krajíček [64] for the completeness of our presentation.

Theorem 30.2.2 (Krajíček [64]) *Assume that Statement (R) is true. Then* EF *is not p-bounded.*

Proof: Assume for the sake of a contradiction that EF is p-bounded and hence, in particular, NP = coNP.

Claim 1: *There is a function* $f : \{0,1\}^* \rightarrow \{0,1\}$ *computable in* NE $\cap$ coNE *such that* $H_f(\ell) \geq 2^{\Omega(\ell)}$.

To see this, note that the property of a string being the lexicographically first string which is a truth table of a function f on $\{0,1\}^\ell$ with $H_f(\ell) \geq 2^{\delta \cdot \ell}$ (some $0 < \delta < 1$) is in the polynomial-time hierarchy PH and the function f that it defines will be in E^{PH}.

But if NP = coNP then such an f is in $E^{\mathrm{NP} \cap \mathrm{coNP}} = \mathrm{NE} \cap \mathrm{coNE}$. This proves the claim.

Having such f we may invoke Statement (R) to deduce that EF is not p-bounded. That is a contradiction. □

We shall see in Chapter 31 (see the remark after the proof of Lemma 31.2.1) that the original requirement from Conjecture 30.2.1 that f is in NP $\cap$ coNP is probably quite important.

We shall prove in Chapter 31 a uniform, bounded arithmetic, version of Conjecture 30.2.1 (i.e. a form of consistency). In fact, the model we construct there is not only a model of PV (a theory corresponding to EF) but even of theory $\mathrm{Th}_{\forall}(L_{\mathrm{PV}})$, the true universal theory in the language of PV, considered in Section 23.1. This theory corresponds in a sense to the class of all proof systems simultaneously (it proves the soundness of all proof systems).

This may cause one to contemplate a strengthening of Conjecture 30.2.1 in that the τ-formulas considered there are hard for all proof systems, not just for EF. We now note that the argument from Section 30.1 (before Possibility A) that the truth-table function is likely not hard for all proof systems can be modified for the NW-generator too, as long as it has an exponential (or at least superpolynomial) number of output bits. This is similar to arguments in Razborov and Rudich [97] and Razborov [96].

Assume, as in (2) above, that m is exponential in n, say $m = 2^{n^\delta}$. Denote $k := n^\delta$, hence $n = k^{O(1)}$. The strings from $\{0,1\}^m$ can be identified with truth tables of Boolean functions with k inputs. The definition of the NW-generator implies that a function $b \in \{0,1\}^m = \{0,1\}^{2^k}$ in its range is computed in a non-uniform NP$\cap$coNP. In particular, if a proof system cannot give a short proof of $\tau(\mathrm{NW}_{A,f})_b$ it also cannot rule out in a short way that b is in non-uniform NP$\cap$coNP.

Arguing as in Section 30.1 it follows that if there is a function in NE$\cap$coNE that is not in non-uniform NP$\cap$coNP then any NW-generator with parameters as in (2) based on an NP$\cap$coNP function (no matter how hard) cannot be hard for all proof systems.

30.3 A possibly hard gadget

We shall discuss a gadget that could possibly lead to a hard generator. A function defined in an analogous way to the NW generator but based on a random sparse $(n+1) \times n$ matrix A and a random function f was proposed as a possibly hard proof complexity generator in Krajíček [63] (see there for the motivation). The qualification 'sparse' means that each row of A contains c 1s where $1 \leq c \leq \log(n)$. A similar construction with an $n \times n$ matrix and c a constant was proposed in Goldreich [37] as a one-way function. We shall now use the idea of gadget generators to remove the randomness part.

The generator $\mathsf{nw}_{n,c}$ will be determined by two parameters $n, c \geq 1$ with $c \leq \log n$. The gadget will have $(n+1)^2$ bits. We shall interpret the first $(n+1)n$ bits as determining the $(n+1) \times n$ matrix A, and denote them accordingly A_{ij}. Then the next 2^c bits of the remaining $n+1$ bits define the truth table of function $f : \{0,1\}^c \to \{0,1\}$, and these bits will be denoted f_ϵ, where $\epsilon \in \{0,1\}^c$. The remaining bits of the gadget are not used (we prefer simple terms to optimal numbers). Of course, we could have encoded A by $O(nc \log n) \leq O(n(\log n)^2)$ bits but at the expense of a less transparent formula defining the generator below.

The function $\mathsf{nw}_{n,c}$ sends k bits to $k+1$ bits where

$$k := (n+1)^2 + n((n+1)^2 + 1).$$

An input $x \in \{0,1\}^k$ is interpreted as a tuple

$$x = (A, f, u^1, \ldots, u^t)$$

where (A, f) is the gadget ($(n+1)^2$ bits) and $u^i \in \{0,1\}^n$ for $i = 1, \ldots, t = (n+1)^2 + 1$.

The value $\mathsf{nw}_{n,c}(x)$ is defined to be the string $0^{(k+1)}$ if the following condition fails:

(1) *A has exactly c ones in each row.*

Otherwise $\mathsf{nw}_{n,c}(x) = y$, where y is a tuple

$$y = (v^1, \ldots, v^t)$$

with

$$\mathrm{NW}_{A,f}(u^i) = v^i, \quad \text{for } i \in [t].$$

Note that $\mathsf{nw}_{n,c}$ is computed by a constant-depth formula of size $O(n^{2+c})$: matrices A satisfying condition (1) are defined by a CNF of size $O(n^{2+c})$, and the ith bit of $\mathrm{NW}_{A,f}(u)$ is defined, assuming (1), by a DNF formula

$$\bigvee_{J \subseteq [n], |J|=c} \left[(\bigwedge_{j \in J} A_{ij}) \wedge (\bigwedge_{\epsilon} (x(J)^{\epsilon} \to f_{\epsilon})) \right]$$

where $\epsilon = (\epsilon_1, \ldots, \epsilon_c)$ ranges over $\{0,1\}^c$, and for $J = \{j_1 < \cdots < j_c\}$

$$x(J)^{\epsilon} := x_{j_1}^{\epsilon_1} \wedge \cdots \wedge x_{j_c}^{\epsilon_c}$$

with $x^1 := x$ and $x^0 := \neg x$. This formula has size $O(c 2^c n^c) \leq O(n^{2+c})$ as $c \leq \log n$.

The following statement appears to be consistent with our present knowledge.

Possibility D: *Function $\mathsf{nw}_{n,c}$, with $c = \log n$, is hard for any proof system.*

Note that the argument from the end of Section 30.2 that the NW generator is probably not hard for all proof systems does not seem to apply here.

It is easy to see that the PHP gadget generator from Section 29.5 is a special case of $\mathsf{nw}_{n,c}$ with $c = 1$ and can be easily embedded if $c > 1$. Hence we have, by Theorem 29.5.2:

Theorem 30.3.1 *For any $1 \leq c \leq \log n$, generator $\mathsf{nw}_{n,c}$ is exponentially hard for constant-depth Frege systems F_d.*

30.4 Rudich's demi-bit conjecture

Let g be a P/poly proof complexity generator. It follows from Lemma 29.1.3 that g is hard for all propositional proof systems iff the complement of the range of g contains no infinite NP set.

A statement related to this property has been put forward as a conjecture by Rudich [99] in a different context, in connection with his attempt to generalize the concept of the so-called *natural proofs* of Razborov and Rudich [97] (see Rudich [99] for this background).

Following Rudich [99] we define the following notion of the hardness of a function. Let h be map $h : \{0,1\}^n \to \{0,1\}^m$, with $m > n$. The **demi-hardness** of h is the minimal s such that there is a non-deterministic circuit C of size at most s which defines a subset C^* of $\{0,1\}^m \setminus \mathrm{Rng}(h)$ of measure $|C^*|/2^m \geq 1/s$.

Conjecture 30.4.1 (Rudich [99, demi-bit conjecture]) *There exists a* P/poly *map* $g = (g_n)_n$, $g_n : \{0,1\}^n \to \{0,1\}^{n+1}$, *and* $\epsilon > 0$ *such that the demi-hardness of* g_n *is at least* 2^{n^ϵ}.

Rudich [99] proposes the subset sum generator of Impagliazzo and Naor [45] as a candidate function to satisfy the conjecture (even a stronger statement, the super-bit conjecture; see Rudich [99]). This generator is defined as follows.

The function g takes as an input k numbers $a_1, \ldots, a_k$ each having $k+1$ bits, and k single bits $b_1, \ldots, b_k$. The output of g is the k-tuple $a_1, \ldots, a_k$ together with

$$\sum_i a_i b_i \ (\mathrm{mod}\ 2^{k+1}),$$

for another $k+1$ bits. Function g thus stretches $n = k(k+2)$ bits into $n+1$ bits.

The generator has several very nice properties that are interesting from the point of view of cryptography (see Impagliazzo and Naor [45]) but Rudich [99] does not offer any particular statement supporting the suggestion that it has also exponential demi-hardness (or even the stronger super-hardness).

31
The local witness model

Our general aim is to construct models relevant to some of the conjectures discussed in Chapter 30, and eventually models that would prove some of them. From a logical point of view this should be more accessible for Razborov's Conjecture 30.2.1 than for the other conjectures as it concerns a function of the highest complexity. This translates into a higher quantifier complexity of the associated formalized statement and thus offers, in principle, more chances to manipulate things to our advantage.

The model we shall construct in this chapter, the local witness model, will yield the following statement. We shall formulate it now a little informally; the formal version in Section 31.4 will quantify the parameters involved.

Theorem 31.0.2 (informal) *Let A be an $m \times n$ 0–1 matrix that is a* $(\log m, n^{1/3})$ *design in the sense of Nisan and Wigderson [84], and $n < m$. Let f be a Boolean function in $n^{1/3}$ variables that is a hard bit of a one-way permutation. Assume R is an infinite* NP*-set.*

Then it is consistent with Cook's theory PV *(and, in fact, with the true universal theory* $\mathrm{Th}_\forall(L_{\mathrm{PV}})$ *in the language of* PV*) that*

$$\mathrm{Rng}(\mathrm{NW}_{A,f}) \cap R \neq \emptyset.$$

This gives a form of consistency of Razborov's conjecture (in fact, of a stronger statement) and also of Rudich's demi-bit conjecture. We shall discuss this in Section 31.4.

31.1 The local witness model $K(F_b)$

We shall continue using the notation from Section 29.4. In particular, $A \in \mathcal{M}_n$ will be an $m \times n$ 0–1 matrix with ℓ 1s per row, in row i the 1s are in columns from

$J_i \subseteq [n]$. We shall fix $\ell := n^{1/3}$ and we shall assume that A is $(\log m, \ell)$-design. We assume that $m > n$ and note that $A \in \mathcal{M}_n$ implies that $m < 2^{n^{1/t}}$, for some $t > \mathbf{N}$ (i.e. m is subexponential in n).

For any $a \in \{0,1\}^n$ and $J = \{j_1 < \cdots < j_\ell\} \subseteq [n]$ denote

$$a(J) := (a_{j_1}, \ldots, a_{j_\ell}).$$

Further we have a function $f : \{0,1\}^\ell \to \{0,1\}$ of the form

$$f(u) := B(h^{(-1)}(u))$$

where h is a polynomial-time permutation and B is a Boolean function. It is intended that h will be a one-way permutation and B its hard bit although that is not needed for the definition of the model (but it is key for its properties).

The model will be parameterized by an element $b \in \mathcal{M}_n$, a string $b = (b_1, \ldots, b_m) \in \{0,1\}^m$, that satisfies

$$b \notin \mathrm{Rng}(\mathrm{NW}_{A,f}).$$

Given such a string, define the **sample space** Ω_b to be the set of all $(m+1)$-tuples

$$\omega = (a, v^1, \ldots, v^m),$$

where $a \in \{0,1\}^n$ and all $v^i \in \{0,1\}^\ell$, and ω satisfies the following condition:

- *For every $i \in [m]$, if $f(a(J_i)) = b_i$ then v^i is the unique string such that*

$$h(v^i) = a(J_i) \wedge B(v^i) = b_i.$$

If no such string exists then $v^i = 0^{(\ell)}$.

We think of v^i as a witness to the fact that the ith bit of $\mathrm{NW}_{A,f}(a)$ has the 'right' value b_i.

From Section 22.5 we know that when aiming at proof complexity we should expect to use partially defined random variables. The **family** F_b of partial random variables forming the local model is the family of all functions $\alpha \in \mathcal{M}_n$, $\alpha : \Omega_b \to \mathcal{M}_n$, satisfying the following condition. There is a non-uniform algorithm S running in time bounded by $2^{n^{1/t}}$, for some $t > \mathbf{N}$ (i.e. in subexponential time) that computes α on a sample $\omega = (a, v^1, \ldots, v^m)$ as follows:

1. *S can read from the input a but not any of v^is.*

2. *S can learn the value of v^i if it queries the parameter i. Upon such a query S is provided with the string v^i, and if*

$$h(v^i) = a(J_i) \wedge B(v^i) = b_i$$

it continues with the computation.
If v^i does not satisfy this condition the computation is aborted and $\alpha(\omega)$ is undefined.
3. *There exists a standard constant $c \geq 1$ such that on every input S queries at most c values of v^i.*
(We could have allowed c up to any number bounded above by $n^{1/t}$, some $t > \mathbf{N}$.)

Note that this type of computation is essentially a computation with an oracle access to a counter-example function in a similar way to Section 24.2 but allowing a partially defined counter-example function. The present formalism makes the 'sources of undefinability' of α more transparent and more amenable to analysis.

We want to show that PV and, in fact, the true universal theory in its language $\mathrm{Th}_{\forall}(L_{\mathrm{PV}})$ (see Section 23.1) is valid in $K(F_b)$ and that

$$(*) \qquad [\![\exists x \mathrm{NW}_{A,f}(x) \;=\; b]\!] = 1_{\mathcal{B}}.$$

However, as we are dealing with only partially defined random variables we have to establish first an important property that each random variable from the family F_b is undefined only on an infinitesimal fraction of the sample space. This fact also plays a crucial role in establishing the equality $(*)$. This is analogous to Lemma 20.1.3 and it is the key property of F_b. In order to establish this fact we shall use the assumed hardness of the function f.

31.2 Regions of undefinability

Recall the definition of the hardness of a function, $H_f(\ell)$, from Section 30.2.1; it is the minimal s such that some circuit of size s computes f with the advantage over $1/2$ of at least $1/s$. Using this concept we can formulate the following key lemma.

Lemma 31.2.1 *Assume that the permutation h and the hard-bit predicate B are such that the function $f := B \circ h^{(-1)}$ they determine has the hardness*

$$H_f(\ell) \geq 2^{\ell^{\epsilon}}$$

for some constant $\epsilon > 0$. *Let* $\alpha \in F_b$ *be arbitrary. Then the probability*

$$\text{Prob}_{\omega \in \Omega_b}[\alpha \text{ is undefined }]$$

is infinitesimal.

The rest of the section is devoted to the proof of this lemma.

Assume that we have a random variable α for which the probability in the lemma is at least (a standard positive) $\delta > 0$. Denote by $W \subseteq \{0,1\}^n$ the set of all $a \in \{0,1\}^n$ such that α is undefined on the unique sample starting with a (the sample is unique because the witnesses are unique). We shall denote such a sample $\omega(a)$, and the witnesses v^i appearing in it $v^i(a)$.

Let S be the algorithm computing α and making on each sample at most c queries about witnesses, as required in Section 31.1. If S aborted the computation on $\omega(a)$ and i was its last query about a witness it means that

$$f(a(J_i)) \neq b_i.$$

Hence we may interpret S as a non-uniform and interactive algorithm solving the following:

Task (T): *Given* $a \in \{0,1\}^n$ *find* $i \in [m]$ *such that the i-th bit of* $\text{NW}_{A,f}(a)$ *differs from* b_i.

By the interactive part of S we mean the ability to ask for a witness and get the correct one, if it exists. To be able to speak about such computations more easily we shall say that 'S asks the oracle for a witness' and that 'the oracle sends a witness'. In this terminology we can reformulate the behavior of S as follows:

- *S, upon receiving an input* $a \in \{0,1\}^n$, *computes his first candidate solution* $i_1 \in [m]$ *for task (T).*
- *If* i_1 *solves (T) (this can be tested in polynomial time) the computation stops.*
- *If* i_1 *fails to solve (T) the oracle sends to S the witness* $v^{i_1}(a)$, *witnessing that* $f(a(J_{i_1})) = b_{i_1}$.
- *In the k-th step S computes a candidate solution* $i_k \in [m]$ *from a and from the* $(k-1)$ *witnesses received in earlier rounds.*
- *If* i_k *solves (T) the computation stops. Otherwise the oracle sends the correct witness* $v^{i_k}(a)$, *and the computation proceeds.*

Assume that for a string $a \in W$ on which S succeeds in solving (T), S asks k queries and then the computation stops: the candidate solutions that S produced were $i_1, \ldots, i_k$ and the last one i_k was correct. We shall call the k-tuple $(i_1, \ldots, i_k)$

the **trace of the computation** on a. In particular, $k \le c$. As the witnesses are unique the trace also determines the oracle replies.

Claim 1: *There is a k-tuple $\mathbf{i} = (i_1, \dots, i_k) \in [m]^k$ for some $k \le c$ that is the trace of computations on at least a fraction of $2/(3m)^k$ of all inputs from W.*

Construct by induction on t a string $(i_1, \dots, i_t) \in [m]^t$ such that the traces of at least $1/3^{t-1}m^t$ fraction of W start with this t-tuple. For $t = 1$ it is obvious.

For the induction step assume that we have constructed a t-tuple $(i_1, \dots, i_t)$ satisfying the property. If $(i_1, \dots, i_t)$ is the complete trace of at least $2/3$ of all inputs from W whose traces start with $(i_1, \dots, i_t)$, it is already the wanted trace.

If it is not the case extend $(i_1, \dots, i_t)$ by i_{t+1} so that for at least $1/(3m)$ of all inputs from W with traces starting with $(i_1, \dots, i_t)$ the traces are not complete and will continue with $(i_1, \dots, i_{t+1})$. It is possible to find such an i_{t+1} as at least a third of all computations with traces starting with $(i_1, \dots, i_t)$ continue and there are m choices for i_{t+1}. This proves the claim.

Let us fix a trace $\mathbf{i} = (i_1, \dots, i_k)$ satisfying the claim. For $u \in \{0,1\}^\ell$ and $v \in \{0,1\}^{n-\ell}$ define string $a(u,v) \in \{0,1\}^n$ by putting bits of u into positions J_{i_k} (in the natural order) and then fill the remaining $n - \ell$ positions by bits of v. The following claim follows by averaging.

Claim 2: *There is an $n - \ell$-tuple $e \in \{0,1\}^{n-\ell}$ such that for a fraction of at least $\delta \cdot \frac{2}{(3m)^k}$ of all $u \in \{0,1\}^\ell$ the string $a(u,e)$ is in W and its trace is* $\mathbf{i}$.

Fix one such an $(n-\ell)$-tuple e. The design property of A implies that there are, for any row $i \ne i_k$, at most $\log(m)$ bits from J_i not set by e. Thus there are at most m assignments w to bits in J_i not set by e. Any such w determines, together with e, an assignment to variables in J_i and hence a string in $\{0,1\}^\ell$; denote it z_w. Let Y_i be the set of all witnesses to

$$f(z_w) = b_i$$

for all such z_w (whenever a witness exists). The total bit size of each Y_i is $m \cdot \ell^{O(1)} = m^{O(1)}$, and there are $m - 1$ of these sets, hence all this information has $m^{O(1)}$ bits.

Define an algorithm C that uses as advice the trace $\mathbf{i}$, the partial assignment e, and all $m - 1$ sets Y_i. The total size of the advice is bounded above by

$$\log(m) \cdot c + (n - \ell) + m^{O(1)} \le m^{O(1)}.$$

The aim of the algorithm is to compute the function f. Let U be those inputs u for which the trace of $a(u,e))$ either equals $\mathbf{i}$ or starts with $\mathbf{i}$, and let b_0 be the majority value of f on the complement of U.

On input $u \in \{0,1\}^\ell$ C defines the string $a := a(u,e) \in \{0,1\}^n$ and starts the computation of S on a. If the first candidate solution that S computes is different from i_1 then C halts and produces b_0 as the output. If the first candidate solution is i_1 the algorithm C reads from Y_{i_1} the right witness y^{i_1}. The uniqueness of witnesses for f implies that there is exactly one suitable string in Y_{i1} and C can find it by testing the condition

$$h(y^{i_1}) = a(J_{i_1}) \wedge B(y^{i_1}) = b_{i_1}.$$

C proceeds analogously for steps $2,\ldots,k-1$: if any candidate solution computed is different from the particular element i_j in the trace, $j = 2,\ldots,k-1$, or if the computation of S halts before reaching the kth step, C halts and outputs the value b_0. Otherwise, in each step, C retrieves from the sets Y_i the correct witness and proceeds with the simulation of S.

Assume that the computation evolved according to the trace $\mathbf{i}$ and C reaches the kth step. If the kth candidate solution is different from i_k, C again outputs b_0. But if it is i_k it outputs $1 - b_{i_k}$.

Claim 3: *The algorithm C computes f correctly on at least a fraction of*

$$1/2 + 2^{-\ell^\epsilon}$$

of inputs, and it is computed by a circuit of size at most 2^{ℓ^ϵ}.

The algorithm C outputs the bit b_0 in all cases except when the computation reaches the kth step and the simulation of S produces i_k as its candidate solution. If the computation of S were to actually stop at that point then the value $1 - b_{i_k}$ would indeed be equal to $f(u)$. If the computation were to continue then we would have no information. But note that by the choice of $\mathbf{i}$ in Claim 1 the former case happens with probability at least $2/3$ and hence for at least a fraction $\delta \cdot \frac{2}{(3m)^k}$ of all inputs $u \in \{0,1\}^\ell$. Because b_0 is the correct value of f for at least half of $u \notin U$, the overall advantage that algorithm C has in computing f is at least $\delta \cdot \frac{1}{(3m)^k}$.

Now note that $m < 2^{n^{1/t}}$ for some $t > \mathbf{N}$, and hence

$$\delta \cdot \frac{1}{(3m)^k} > 2^{-\ell^\epsilon}$$

as $\epsilon > 0$ is standard.

The algorithm C uses advice of size $m^{O(1)}$ and needs the same time as S except when it needs to simulate a reply of the oracle and to find an appropriate witness in one of the sets Y_i. This is done at most $(c-1)$-times and takes $m^{O(1)}$ time each. Hence the total time is $m^{O(1)}$ which is also less than 2^{ℓ^ϵ}.

This establishes the claim.

Claim 3 contradicts the assumed hardness of f and hence the lemma follows. □

Let us make an observation relevant to Theorem 30.2.2. Assume we would want to allow f not only from $NP \cap coNP$ but from $NE \cap coNE$. In such a case each step in which C simulates the oracle and searches for a witness would take time $2^{O(\ell)}$. Hence the size of C would be estimated by $O(m^{O(1)}2^{O\ell})$, and one gets no contradiction with the hardness of f.

31.3 Properties of the local model $K(F_b)$

Having the key Lemma 31.2.1 we are in a position to state and prove the required properties of $K(F_b)$. Let A, f and b be as in Section 31.1.

Theorem 31.3.1 *Assume the function f has an exponential hardness*

$$H_f(\ell) \geq 2^{\ell^\epsilon}$$

for some $\epsilon > 0$.

Then the true universal theory in the language of PV, *theory* $\mathrm{Th}_\forall(L_{PV})$, *is valid in $K(F_b)$, and so is therefore theory* PV. *Also*

$$[\![\exists x \mathrm{NW}_{A,f}(x) = b]\!] = 1_{\mathcal{B}}.$$

Proof: $K(F_b)$ is clearly L_{PV}-closed and hence the validity of $\mathrm{Th}_\forall(L_{PV})$ follows from the general Lemma 1.4.2.

To compute the truth value of the statement $\exists x \mathrm{NW}_{A,f}(x) = b$ write it in full as

$$\exists x\ \forall i (i \in [m])\ \exists y (|y| = \ell)\ B(x, i, y),$$

where $B(x,i,y)$ is the open PV-formula (with parameters A, b)

$$h(y) = x(J_i)\ \wedge\ B(y) = b_i.$$

Let $\alpha \in F_b$ be the function that is just the projection on the first coordinate of the sample:

$$\alpha((a, v^1, \ldots, v^m)) := a.$$

We want to show that

$$[\![\ \forall i(i \in [m])\ \exists y(|y| = \ell)\ B(\alpha, i, y)\ ]\!] \ = \ 1_{\mathcal{B}}.$$

For this it is necessary to show that for each $\iota \in F_b$

$$[\![\ \iota \in [m] \ \rightarrow \ \exists y(|y| = \ell)\ B(\alpha, \iota, y)\ ]\!] \ = \ 1_{\mathcal{B}}.$$

Given ι take random variable $\gamma \in F_b$ defined on a sample $\omega = (a, v^1, \ldots, v^m)$ as follows:

$$\gamma(\omega) = \begin{cases} v^i & \text{if } \iota(\omega) = i \in [m] \wedge h(v^i) = a(J_i) \wedge B(v^i) = b_i \\ \text{undefined} & \text{otherwise.} \end{cases}$$

The random variable γ can be clearly computed in the manner required for F_b, and by the key Lemma 31.2.1 it is undefined only at an infinitesimal fraction of samples. Hence

$$\mathrm{Prob}_{\omega \in \Omega_b}[B(\alpha(\omega), \iota(\omega), \gamma(\omega))]$$

is infinitesimally close to 1, and the equality

$$[\![B(\alpha(\omega), \iota(\omega), \gamma(\omega))]\!] \ = \ 1_{\mathcal{B}}$$

is established. □

31.4 What does and what does not follow from $K(F_b)$

We shall now reformulate Theorem 31.0.2 formally, specifying the requirements on the parameters involved, and use Theorem 31.3.1 to prove it. This statement does not use any non-standard objects.

Theorem 31.4.1 *Let $k \geq 1$ be a natural number parameter and $m = m(k)$ a function of k such that*

$$k \ < \ m(k) \ < \ 2^{k^{o(1)}}.$$

Let A_k be $k \times m$ 0–1 matrices that are $(\log m, k^{1/3})$ designs. Let f be an NP ∩ coNP *function that is a hard bit of a one-way permutation, and assume that it has an exponential hardness*

$$H_f(\ell) \ \geq \ 2^{\ell^{\epsilon}}, \quad \textit{for some } \epsilon > 0.$$

Let $g : \{0,1\}^* \rightarrow \{0,1\}^*$ *be the map that is defined on* $\{0,1\}^k$ *as* $\mathrm{NW}_{A_k,f}$.

Assume that R is an infinite NP *set and that it satisfies the following lengths-condition:*

- *There are infinitely many elements of R whose length equals* $m(k)$ *for some* $k \geq 1$.

Then it is consistent with Cook's theory PV *and, in fact, with the true universal theory* $\mathrm{Th}_\forall(L_{\mathrm{PV}})$ *in the language of* PV, *that*

$$\mathrm{Rng}(\mathrm{NW}_{A,f}) \cap R \neq \emptyset.$$

Proof: Let g be a map and R be an infinite NP set satisfying the hypothesis of the theorem. If R intersects the range of g there is nothing to prove. So assume that R and $\mathrm{Rng}(g)$ are disjoint.

As R is infinite and satisfies the lengths condition it follows by the overspill in $\mathcal{M}$ (see the Appendix) that R has a non-standard element (in fact, there are cofinaly many of them) of some non-standard length m of the form $m = m(n)$ for some non-standard n. Moreover, it is not in the range of g. Fix one such non-standard element b, and parameters n and m.

Take n as the parameter defining the cut $\mathcal{M}_n$. Then $b \in \mathcal{M}_n$ and we may apply Theorem 31.0.2 and construct the Boolean-valued structure $K(F_b)$ in which PV and $\mathrm{Th}_\forall(L_{\mathrm{PV}})$ are valid but also

$$[\![\, b \in \mathrm{Rng}(g) \,]\!] \;=\; 1_{\mathcal{B}}.$$

The membership of b in R is witnessed in $\mathcal{M}$ by a string of length $m^{O(1)}$ and thus this string is in $\mathcal{M}_n$, and hence in $K(F_b)$ too. Hence

$$[\![\, b \in R \,]\!] \;=\; 1_{\mathcal{B}},$$

and we have

$$[\![\mathrm{Rng}(g) \cap R \;\neq\; \emptyset \,]\!] \;=\; 0_{\mathcal{B}}.$$

The theorem follows from Lemma 1.4.1. □

For the sake of an easier discussion to follow (and to avoid confusion) let us formulate the statement from the theorem whose consistency was shown (for each R) as a statement of its own.

Statement (S): *Let $k \geq 1$ be a natural number parameter and $m = m(k)$ a function of k such that*

$$k < m(k) < 2^{k^{o(1)}}.$$

Let A_k be $k \times m$ 0–1 matrices that are $(\log m, k^{1/3})$ designs. Let f be an NP$\cap$coNP *function that is a hard bit of a one-way permutation, and assume that it has an exponential hardness*

$$H_f(\ell) \geq 2^{\ell^\epsilon}, \quad \text{for some } \epsilon > 0.$$

Let $g : \{0,1\}^* \to \{0,1\}^*$ be the map that is defined on $\{0,1\}^k$ as $\mathrm{NW}_{A_k,f}$*. Assume that R is an infinite* NP*-set which has infinitely many elements whose length equals $m(k)$ for some $k \geq 1$.*

Then

$$\mathrm{Rng}(NW_{A,f}) \cap R \neq \emptyset.$$

Lemma 31.4.2 *Statement (S) implies Razborov's conjecture 30.2.1 for all proof systems* P *(not just* EF*).*

Proof: The NW generator from the conjecture satisfies the properties of the map g required in Statement (S).

Let P be a proof system and $k \geq 1$ a constant. The set of all b for which the τ-formula $\tau(g)_b$ has a P-proof of length at most $|\tau(g)_b|^k$ is in NP and it is disjoint (by the soundness of P) from the range of g. It also trivially satisfies the lengths condition.

That is a contradiction with Statement (S). □

We thus view Theorem 31.4.1 as relevant to Conjecture 30.2.1. In addition, theory PV corresponds to EF, and theory $\mathrm{Th}_\forall(L_{PV})$ corresponds to (the union of) all proof systems.

Theorem 31.4.1 seems to be also relevant (or, at least, related) to Rudich's demi-bit Conjecture 30.4.1. Take Statement (S) for $m := n+1$. In that case the lengths condition is void. The cut $\mathcal{M}_n$ contains all numbers of length subexponential in n, hence the theorem yields (as $m = n+1$) the consistency of Statement (S) with R from subexponential non-deterministic time as in Conjecture 30.4.1 (but the conjecture allows also non-uniform algorithms).

Statement (S) is weaker than Conjecture 30.4.1 in the sense that the latter claims the existence of a P/poly map while in (S) we have an NP$\cap$coNP-map. On the other hand the conjecture is weaker as it does not rule out infinite sets R but only those having at least a subexponential measure.

Let us conclude this chapter by spelling out what does not follow from the construction. Assume for a moment that Conjecture 30.2.1 is not true or, more generally, that Statement (S) is not true. Let f be a hard NP$\cap$coNP-function and denote by g the NW generator based on it (with the parameters as in this chapter). Assume that g is not hard for a proof system P. By the overspill in $\mathcal{M}$ we get a non-standard n, a string $b \in \mathcal{M}_n$ of length $m = m(n)$ that is not in the range of g, and also a P-proof $\pi \in \mathcal{M}_n$ of $\tau(g)_b$. The proof π will be present also in $K(F_b)$ and the soundness of P is valid there (the soundness is a true universal statement). But the statement that the τ-formula expresses,

$$\forall x \exists i \in [m] \forall y^i \neg[h(y^i) = x(J_i) \wedge B(y^i) = b_i],$$

and π proves, is not valid in $K(F_b)$.

The reader may wonder why in this situation we cannot apply the approach of Chapter 18 (see also Theorem 28.2.1) and derive a contradiction, thus proving Statement (S). The reason is as follows. The variables of the τ-formula correspond to (bits of) x and of all y^i. In the model $K(F_b)$ we have a truth assignment corresponding to a witness for $\exists x$ in

$$\exists x \forall i \in [m] \exists y^i \; [h(y^i) = x(J_i) \wedge B(y^i) = b_i]$$

and also truth assignments corresponding to individual witnesses y^i but we do not have a string there that would collect all these assignments together. The principle asserting the existence of such a string is the so-called sharply bounded collection scheme and, unfortunately, it is not provable in $Th_\forall(L_{PV})$ unless factoring is easy (see Cook and Thapen [32]). But such a string seems to be needed in order to apply the soundness of P.

What we need to arrange in the model is the following form of a reflection principle for P. Denote by $\varphi_i(y^i)$ the propositional translation of

$$\neg[h(y^i) = \alpha(J_i) \wedge B(y^i) = b_i],$$

a formula defined in $K(F_b)$. The substitution instance of π (substituting for x) is a proof of a disjunction

$$\bigvee_i \varphi_i(y^i)$$

in disjoint sets of variables. For each φ_i we have a falsifying assignment. This situation should not be possible in the constructed model of $Th_\forall(L_{PV})$.

The reader may still wonder if the construction cannot be modified to achieve this. I am wondering too; Krajíček [72] presents (in classical terms) variants of the construction, considers quantitative aspects of relations between various parameters, and discusses possibilities of how to extend the construction. Perhaps the most promising avenue is to try to employ the **feasible disjunction property** which can be assumed to hold, for the purpose of proving lower bounds, without loss of generality for every proof system (see Krajíček [72]).

Appendix: Non-standard models and the ultrapower construction

In this Appendix we present the ultrapower construction of a non-standard model of the true arithmetic. The construction is quite elementary and intuitive, and we present it in a self-contained manner (proofs of all statements are included). Indeed, we also discuss a few other basic notions and terminology useful when dealing with non-standard models. All this material is included to help a reader with less experience in mathematical logic to create a mental image of model $\mathcal{M}$ that forms the ambient universe for our constructions (Section 1.1).

For the presentation of the construction we fix the language to be L_{PA}, the language of Peano arithmetic (Section 5.1), having constants 0 and 1, two binary functions $x+y$ and $x \cdot y$, and the binary relation $x < y$. When talking about arithmetic in mathematical logic this is the usual choice of language. In Section 1.1 we opted for a much larger language L_{all} as the language of $\mathcal{M}$; we shall discuss the differences after Theorem A.9 but for now we want to keep the whole picture as elementary as possible.

Each symbol from L_{PA} has a canonical interpretation in the set of natural numbers $\mathbf{N}$. We shall denote the resulting L_{PA}-structure also $\mathbf{N}$. The phrase **the true arithmetic** means $\mathrm{Th}_{L_{\mathrm{PA}}}(\mathbf{N})$, the L_{PA}-theory of $\mathbf{N}$, i.e. the set of all L_{PA}-sentences true in $\mathbf{N}$. The structure $\mathbf{N}$ is called the **standard model** of arithmetic.

Several general theorems of mathematical logic guarantee the existence of L_{PA}-structures that satisfy the same theory as $\mathbf{N}$, i.e. they are **elementary equivalent** to $\mathbf{N}$, but are not isomorphic to $\mathbf{N}$. Such structures are called **non-standard models** of true arithmetic. We shall later employ the ultrapower construction to describe one such model. But before we do so we shall observe a few basic properties shared by all non-standard models of arithmetic.

Let $\mathbf{M}$ denote a non-standard model $(M, 0^{\mathbf{M}}, 1^{\mathbf{M}}, =^{\mathbf{M}}, \cdot^{\mathbf{M}}, <^{\mathbf{M}})$ of true arithmetic. It holds that $<$ is a strict linear ordering in $\mathbf{N}$ and this can be expressed by one L_{PA}-sentence, hence the same is true in $\mathbf{M}$. Analogous arguments yield all the following statements. Constant 0 is $<$-minimal in $\mathbf{N}$, hence $0^{\mathbf{M}}$ is $<^{\mathbf{M}}$-minimal in $\mathbf{M}$ too. Constant 1 is bigger than 0 in $\mathbf{N}$ but there is no element between them, and thus $1^{\mathbf{M}}$ is bigger than $0^{\mathbf{M}}$ in $\mathbf{M}$, with no element of $\mathbf{M}$ in between.

Each positive element $0 < k \in \mathbf{N}$ is the value of a closed L_{PA} term s_k of the form

$$s_k := 1+(1+(1+\ldots(1+1)\ldots)$$

with k occurrences of 1. Put also $s_0 := 0$. In $\mathbf{N}$ it is true for each such term s_k that s_k is smaller than s_{k+1} without an element in between. Hence again this will hold in $\mathbf{M}$ too about values $(s_k)^{\mathbf{M}}$ of the terms in the structure. Furthermore, for any $k, \ell \in \mathbf{N}$ it holds in $\mathbf{N}$ that:

$$s_k + s_\ell = s_{k+\ell} \quad \text{and} \quad s_k \cdot s_\ell = s_{k\cdot\ell}.$$

These L_{PA}-sentences are valid in $\mathbf{M}$ too.

Consequently, the sequence of values of L_{PA}-terms in $\mathbf{M}$:

$$0^{\mathbf{M}}, 1^{\mathbf{M}}, (s_2)^{\mathbf{M}}, (s_3)^{\mathbf{M}}, \ldots$$

forms an initial segment of $\mathbf{M}$ (with respect to the ordering $<^{\mathbf{M}}$) that is isomorphic to $\mathbf{N}$. They are called **standard numbers**. It is customary (in order to simplify the notation) to identify $\mathbf{N}$ with this initial segment and to denote $(s_k)^{\mathbf{M}}$ simply k; hence we think that $\mathbf{N} \subseteq \mathbf{M}$.

As we have assumed that $\mathbf{N}$ and $\mathbf{M}$ are not isomorphic there must be an element $n \in \mathbf{M} \setminus \mathbf{N}$. Each such element is called a **non-standard number**. As there can be nothing between the values of s_k and s_{k+1}, n is bigger than all standard numbers k.

Once we have one non-standard number n we can argue for the existence of many more. For example, $n+1, n+2, \ldots$ are non-standard. But there must also be an element m such that $m+1 = n$ (as that is true in $\mathbf{N}$ about all numbers bigger than 0, which n is in $\mathbf{M}$). It is sensible to denote this $n-1$. And there will be $n-2 = (n-1)-1, n-3, \ldots$. Thus each non-standard number has around itself a block of numbers that is order-isomorphic to the integers $\mathbf{Z}$.

There will be also such a $\mathbf{Z}$-block around $n+n$ and it is easy to see that they will be disjoint as there are infinitely many elements of $\mathbf{M}$ between n and $n+n$. In the other direction there must be a unique element m such that $m+m = n$ or $(m+m)+1 = n$; denote it $\lfloor \frac{n}{2} \rfloor$. The $\mathbf{Z}$-block around $\lfloor \frac{n}{2} \rfloor$ is again disjoint from the **Z**-block around n. Further there will be elements with values $\lfloor \frac{3n}{4} \rfloor$ and $\lfloor \frac{3n}{2} \rfloor$, etc. The reader can verify in this elementary manner that the **Z**-blocks of non-standard numbers are linearly ordered by $<^{\mathbf{M}}$ and that this ordering is dense, and has neither minimal nor maximal block.

Speaking of integers: integers **Z** and rational numbers **Q** are easy to construct from natural numbers **N** and the same construction applied to a non-standard model **M** will give non-standard integers and rationals. We called the latter **M**-rationals in Section 1.1. An **M**-rational q that is positive but smaller than $\frac{1}{k}$ for any standard $k \in \mathbf{N}$, is called **infinitesimal**. For example, if $n \in \mathbf{M}$ is non-standard then $\frac{1}{n}$ is infinitesimal.

Elementary but very useful properties of non-standard models are the so called **Overspill** and **Underspill**.

Theorem A.1 (Overspill) *Let* **M** *be a non-standard model of true arithmetic and let* $\varphi(x)$ *be any* L_{PA}*-formula possibly with parameters from* **M**.

Assume $\varphi(k)$ *holds in* **M** *for all standard numbers* k*. Then there is also a non-standard number* n *such that* $\varphi(n)$ *holds.*

Proof: Assume $\varphi(x)$ has the form $\psi(x, b_1, \ldots, b_t)$ with $b_1, \ldots, b_t \in M$, where $\psi(x, y_1, \ldots, y_t)$ has no parameters.

Assume for the sake of contradiction that $\varphi(k)$ holds for all standard numbers k but for no non-standard number. Then in **M** the following L_{PA}-sentence holds:

$$\exists y_1, \ldots, y_t \, [\, \psi(0, y_1, \ldots, y_t) \wedge \forall x(\psi(x, y_1, \ldots, y_t) \rightarrow \psi(x+1, y_1, \ldots, y_t))$$

$$\wedge \neg \forall x \psi(x, y_1, \ldots, y_t) \,].$$

But that is impossible as the sentence can never hold in **N**: it would violate induction. □

The reader can verify that analogous reasoning also gives the Underspill.

Theorem A.2 (Underspill) *Let* **M** *be a non-standard model of true arithmetic and let* $\varphi(x)$ *be any* L_{PA}*-formula possibly with parameters from* **M**.

Assume $\varphi(n)$ *holds in* **M** *for arbitrarily small non-standard number* n*: for each non-standard* m *there is a non-standard* $n < m$ *such that* $\varphi(n)$ *holds.*

Then there is also a standard number k *such that* $\varphi(k)$ *holds.*

These two properties are shared by all non-standard models. But the two properties (1) and (2) of $\mathcal{M}$ from Section 1.1 (see also below) are not. To arrange these two properties we have to construct specific models. We now embark on one such construction, the ultrapower.

Let us denote by $\prod_\omega \mathbf{N}$ the set of all functions

$$f : \omega \to \mathbf{N}.$$

Here ω and $\mathbf{N}$ are the same sets but the role of arguments and values is quite different in the construction and we prefer to show this in the notation as well.

The universe of the future model will be 'essentially' formed by $\prod_\omega \mathbf{N}$. The qualification 'essentially' is, however, crucial. For the success of the construction it will be necessary to identify those functions f and g that agree on most but not necessarily all arguments. For example, if f and g are eventually equal, i.e. equal for large enough arguments, then we will identify them. In other words, whether or not two functions f and g denote the same element of the universe will depend only on whether or not the set

$$\langle\langle f = g \rangle\rangle := \{i \in \omega \mid f(i) = g(i)\}$$

is 'large'. Every cofinite set will be large but we need to define the 'largeness' for all subsets of ω. The concept that formalizes this informal notion is the **non-principal ultrafilter**.

Definition A.3 A set $\mathcal{U} \subseteq \mathcal{P}(\omega)$ of some subsets of ω is called a **filter** iff it satisfies the following three conditions:

1. $\emptyset \notin \mathcal{U}$ and $\omega \in \mathcal{U}$.
2. If $I, J \in \mathcal{U}$ then $I \cap J \in \mathcal{U}$.
3. If $I \in \mathcal{U}$ and $J \supseteq I$ then $J \in \mathcal{U}$.

A filter that also satisfies

4. For any $I \subseteq \omega$ either $I \in \mathcal{U}$ or $\bar{I} := \omega \setminus I \in \mathcal{U}$

is an **ultrafilter**.

An ultrafilter that contains no finite set is called **non-principal**.

It is the fourth property that makes the notion of a non-principal ultrafilter quite non-transparent; the first three conditions are satisfied, for example, by the class of cofinite sets. Indeed, there is no similarly explicit example of a non-principal ultrafilter and its existence can be proved only from the Axiom of Choice. We shall leave these set-theoretic issues aside and simply take the existence of a non-principal ultrafilter $\mathcal{U}$ as given. The reader may consult Jech [49] for details.

Armed with $\mathcal{U}$ we define an equivalence relation on $\prod_\omega \mathbf{N}$ as follows:

$$f \sim g \text{ iff } \langle\langle f = g \rangle\rangle \in \mathcal{U}.$$

Lemma A.4 *Relation $\sim$ is an equivalence relation on $\prod_\omega \mathbf{N}$.*

Proof: $\langle\!\langle f = f\rangle\!\rangle = \omega$, which is in $\mathcal{U}$ by condition 1 of Definition A.3, and clearly $\langle\!\langle f = g\rangle\!\rangle = \langle\!\langle g = f\rangle\!\rangle$. Hence $\sim$ is reflexive and symmetric.

Assume $\langle\!\langle f = g\rangle\!\rangle = I \in \mathcal{U}$ and $\langle\!\langle g = h\rangle\!\rangle = J \in \mathcal{U}$. Clearly $\langle\!\langle f = h\rangle\!\rangle \supseteq I \cap J$ and hence $\langle\!\langle f = h\rangle\!\rangle \in \mathcal{U}$ by conditions 2 and 3 of Definition A.3. That is, $\sim$ is transitive. □

For a function $f \in \prod_\omega \mathbf{N}$ denote by $[f]$ the $\sim$-equivalence class of f. The universe of the future model consists of these equivalence classes. We shall denote the universe, as well as the whole L_{PA}-structure on it, as $\mathbf{N}^*$ (the usual notation is $\prod_{\mathcal{U}} \mathbf{N}$).

To make $\mathbf{N}^*$ into a structure we need to interpret on it the symbols from language L_{PA}. We shall denote the interpretations by the superscript*. Let

$$0^* := [c_0] \text{ and } 1^* = [c_1]$$

where c_0 and c_1 are functions on ω that are constantly 0 or 1, respectively. Next define:

$$[f] +^* [g] := [f + g]$$

and

$$[f] \cdot^* [g] := [f \cdot g].$$

Finally:

$$f <^* g \text{ iff } \langle\!\langle f < g\rangle\!\rangle \in \mathcal{U}.$$

Lemma A.5 *The functions $+^*$ and $\cdot^*$ and the relation $<^*$ are well-defined on $\mathbf{N}^*$, i.e. the definitions do not depend on the choice of representants f and g from the equivalence classes $[f]$ and $[g]$.*

Proof: We need to show that if $f \sim f'$ and $g \sim g'$ then

$$[f + g] = [f' + g'] \text{ and } [f \cdot g] = [f' \cdot g']$$

and

$$\langle\!\langle f < g\rangle\!\rangle \in \mathcal{U} \text{ iff } \langle\!\langle f' < g'\rangle\!\rangle \in \mathcal{U}.$$

Denote $I := \langle\!\langle f = f'\rangle\!\rangle$ and $J := \langle\!\langle g = g'\rangle\!\rangle$. Hence $I, J \in \mathcal{U}$.

We have

$$\langle\!\langle f + f' = g + g'\rangle\!\rangle \supseteq I \cap J.$$

As the right-hand side is in $\mathcal{U}$ by condition 2 of Definition A.3, so is the left-hand side, by condition 3. That is, $[f + f'] = [g + g']$. The argument for the multiplication is identical.

To check the property for the ordering relation, denote $A := \langle\langle f < g \rangle\rangle$ and $B := \langle\langle f' < g' \rangle\rangle$, and assume $A \in \mathcal{U}$. As $I, J, A \in \mathcal{U}$, also $I \cap J \cap A \in \mathcal{U}$ (condition 2). But $B \supseteq I \cap J \cap A$ and so $B \in \mathcal{U}$ too (condition 3). The opposite direction is analogous. □

Now we have our structure

$$\mathbf{N}^* = (\mathbf{N}^*, 0^*, 1^*, +^*, \cdot^*, <^*)$$

but we still need to establish that it is a model of true arithmetic, and that it is non-standard. For both tasks we need some criterion of how to decide what is true and what is false in the structure. Such a criterion is provided by a simple and beautiful theorem of Loś. (The notation $\langle\langle \ldots \rangle\rangle$ is not usual in this context; we have borrowed it from our earlier chapters.)

Theorem A.6 (Loś theorem) *Let $\varphi(x_1, \ldots, x_k)$ be an L_{PA} formula with all free variables among $x_1, \ldots, x_k$, and let $f_1, \ldots, f_k \in \prod_\omega \mathbf{N}$ be arbitrary. Denote by*

$$\langle\langle \varphi(f_1, \ldots, f_k) \rangle\rangle$$

the set of all $i \in \omega$ such that

$$\mathbf{N} \models \varphi(f_1(i), \ldots, f_k(i)).$$

Then

$$\mathbf{N}^* \models \varphi([f_1], \ldots, [f_k]) \text{ iff } \langle\langle \varphi(f_1, \ldots, f_k) \rangle\rangle \in \mathcal{U}.$$

Proof: This is proved by induction on the number of connectives and quantifiers in φ. To start the induction we need to establish the theorem for atomic formulas $t = s$ and $t < s$, where t and s are L_{PA}-terms. This follows readily from the following property of terms that is verified by induction on the number of function symbols occurring in the term, using the definition of the operations on $\mathbf{N}^*$.

Claim 1: *Let $t(y_1, \ldots, y_\ell)$ be an L_{PA}-term with variables among $y_1, \ldots, y_\ell$, and let $g_1, \ldots, g_\ell$ be arbitrary functions from $\prod_\omega \mathbf{N}$.*

Then

$$t([g_1], \ldots, [g_\ell]) = [t(g_1, \ldots, g_\ell)].$$

If φ is a Boolean combination of smaller formulas we use for the induction step the following claim, a straightforward consequence of the definition of an ultrafilter.

Claim 2: *For any $I, J \subseteq \omega$:*

1. $I \cap J \in \mathcal{U}$ *iff* $I \in \mathcal{U} \wedge J \in \mathcal{U}$,
2. $I \cup J \in \mathcal{U}$ *iff* $I \in \mathcal{U} \vee J \in \mathcal{U}$,
3. $I \in \mathcal{U}$ *iff* $\bar{I} \notin \mathcal{U}$.

Let us see how this handles the Boolean connectives. Assume for instance that $\varphi = \psi \wedge \eta$. By the induction hypothesis we have the Loś theorem for both ψ and η. Then using Tarski's satisfiability conditions in a structure we compute:

$$\mathbf{N}^* \models \varphi([f_1],\ldots,[f_k])$$

iff

$$\mathbf{N}^* \models \psi([f_1],\ldots,[f_k]) \quad \text{and} \quad \mathbf{N}^* \models \eta([f_1],\ldots,[f_k])$$

iff

$$\langle\langle \psi(f_1,\ldots,f_k)\rangle\rangle \in \mathcal{U} \quad \text{and} \quad \langle\langle \eta(f_1,\ldots,f_k)\rangle\rangle \in \mathcal{U}$$

iff (by condition 1 of Claim 2)

$$\langle\langle \psi(f_1,\ldots,f_k)\rangle\rangle \cap \langle\langle \eta(f_1,\ldots,f_k)\rangle\rangle \in \mathcal{U}$$

iff

$$\langle\langle \varphi(f_1,\ldots,f_k)\rangle\rangle \in \mathcal{U}.$$

If φ is a disjunction or a negation, an analogous argument works, using the other two conditions from Claim 2. Note that the first two conditions hold also for a filter but the third one needs an ultrafilter.

Finally we need to treat the case when φ starts with a quantifier. Let $\varphi = \exists y \rho$ and assume first that

$$\mathbf{N}^* \models \exists y \rho([f_1],\ldots,[f_k],y)$$

and hence

$$\mathbf{N}^* \models \rho([f_1],\ldots,[f_k],[g])$$

for some $g \in \prod_\omega \mathbf{N}$. By the induction hypothesis this is equivalent to

$$\langle\langle \rho(f_1,\ldots,f_k,g)\rangle\rangle \in \mathcal{U}.$$

Clearly

$$\langle\langle \rho(f_1,\ldots,f_k,g)\rangle\rangle \subseteq \langle\langle \exists y \rho(f_1,\ldots,f_k,y)\rangle\rangle$$

and hence

$$\langle\langle \exists y \rho(f_1,\ldots,f_k,y)\rangle\rangle = \langle\langle \varphi(f_1,\ldots,f_k)\rangle\rangle \in \mathcal{U}$$

too. This establishes one direction of the equivalence from Loś theorem.

The other direction is also simple. Assume

$$\langle\langle \exists y \rho(f_1,\ldots,f_k,y)\rangle\rangle \in \mathcal{U}.$$

Take a function g defined as follows: $g(i)$ is equal to the minimal number $j \in \mathbf{N}$ such that

$$\mathbf{N} \models \rho(f_1(i),\ldots,f_k(i),j)$$

if it exists, and to 0 otherwise. Clearly then

$$\langle\langle \exists y \rho(f_1,\ldots,f_k,y)\rangle\rangle = \langle\langle \rho(f_1,\ldots,f_k,g)\rangle\rangle$$

and hence

$$\langle\langle \rho(f_1,\ldots,f_k,g)\rangle\rangle \in \mathcal{U}$$

too. The induction hypothesis then yields

$$\mathbf{N}^* \models \rho([f_1],\ldots,[f_k],[g]),$$

which implies the desired result

$$\mathbf{N}^* \models \exists y \rho([f_1],\ldots,[f_k],y).$$

The case of the universal quantifier is completely analogous (alternatively we can replace it by a combination of the existential one with two negations). □

Note that for φ an L_{PA} sentence without parameters we either have $\langle\langle\varphi\rangle\rangle = \omega$ or $\langle\langle\varphi\rangle\rangle = \emptyset$, the former when φ holds in $\mathbf{N}$ and the latter when it does not. This immediately yields the following statement.

Corollary A.7 $\mathbf{N}^*$ *is a model of true arithmetic.*

So far we have used that $\mathcal{U}$ is an ultrafilter but not that it is non-principal. This is used in the next statement.

Lemma A.8 $\mathbf{N}^*$ *is non-standard.*

Proof: The initial segment in $\mathbf{N}^*$ that is isomorphic to $\mathbf{N}$ is the sequence

$$[c_0],[c_1],[c_2],\ldots$$

where c_k is the constant function with the value $k \in \mathbf{N}$. This is easy to verify using Łoś theorem.

Define function f by:

$$f(i) := i.$$

Then for each k the set $\langle\langle c_k = f \rangle\rangle$ is finite and therefore, by the assumption of non-principality, not in $\mathcal{U}$. Hence by Loś theorem $[f] \neq [c_k]$, for each k. □

It may be illustrative for the reader to define functions of various growth rates and compare the elements of the non-standard model they define. This is quite consistent with the idea of comparing functions by their asymptotic behavior.

It remains to establish for $\mathbf{N}^*$ the two properties (1) and (2) of $\mathcal{M}$ from Section 1.1. The properties formulated for $\mathbf{N}^*$ are:

(1) If a_k, $k \in \mathbf{N}$, is a countable family of elements of $\mathbf{N}^*$ then there exists a non-standard $t \in \mathbf{N}^*$ and a sequence $(b_i)_{i<t} \in \mathbf{N}^*$ such that $b_k = a_k$ for all $k \in \mathbf{N}$.
(2) If A_k, $k \in \mathbf{N}$, is a countable family of definable subsets of $\mathbf{N}^*$ such that $\bigcap_{i<k} A_i \neq \emptyset$ for all $k \geq 1$, then $\bigcap_k A_k \neq \emptyset$.

In property (1) we talk about a sequence of non-standard length as being an element of the model. Formally one needs to show how a single number can code a sequence of other numbers in a way that the ith element of the sequence can be defined (from the code and the index i) by a formula of the arithmetic language. This requires some work (and goes back to Gödel); the reader can consult Hájek and Pudlák [38]. However, there is an informal picture that helps to understand these statements even to those unfamiliar with the development of formal arithmetic. This is based on the observation (which, when proved in detail, is as technical as the coding) that instead of numbers we can think of finite sets. It is well-known how the set universe interprets natural numbers (and then goes on interpreting ordinals etc.) and in this sense we can interpret natural numbers in the universe of hereditary finite sets. But there is also an interpretation going in the opposite direction. For example, think of 0 as coding the empty set $\emptyset$ and a number $k = 2^{u_1} + \ldots 2^{u_t} > 0$ with $u_1 > \cdots > u_t \geq 0$ as coding the set whose elements are the sets coded by $u_1, \ldots, u_t$.

Everybody (we hope) learns in school how pairs, sequences, functions, relations, etc. are represented by sets. Hence the interpretation of sets by numbers allows us to transfer the same representation to the universe of numbers.

When the universe of numbers (i.e. a model of true arithmetic) contains also non-standard numbers it happens that some of them will code sequences whose length, while finite in the sense of the model (that is, the length is a number from the universe), will be non-standard. Similarly, a set coded in a non-standard model $\mathbf{M}$ may be finite but it may also be infinite. In the latter case its cardinality in the model is counted by some (necessarily non-standard) number of the model. Sets of both type are called $\mathbf{M}$-finite.

An **M**-finite set is definable in the model but the opposite is not true; for example, the set of odd numbers is obviously definable but not **M**-finite. It is easy to verify by induction that a definable set is coded by a number, is **M**-finite, iff it is **bounded** in **M**: there is a number in **M** bounding above all elements of the set.

I hope this helps the reader not familiar with the topic of formal arithmetic to understand the statements and the construction below. However, there is a limit to how far one can travel with informal mental images. At some point it may be necessary to read a more formal treatment in a textbook.

Theorem A.9 *Both properties (1) and (2) hold for* $\mathbf{N}^*$.

Proof: First observe that (1) follows from (2). To see this define in $\mathbf{N}^*$ sets A_k as follows: A_k contains all sequences (tacitly, $\mathbf{N}^*$-finite and coded in $\mathbf{N}^*$) that start with an initial segment $(a_0, a_2, \ldots, a_k)$.

Clearly $(a_0, a_2, \ldots, a_{k-1}) \in \bigcap_{i<k} A_i$ and hence the intersection is non-empty. Condition (2) then implies that the intersection of all sets A_k must be non-empty. In other words, there must be a sequence whose length is a non-standard number and whose kth element is a_k, for all standard k.

To prove property (2) assume that A_k are definable subsets of $\mathbf{N}^*$ with non-empty finite intersections. Replacing A_k by $\bigcap_{i\leq k} A_i$ we may assume that

$$A_0 \supseteq A_1 \supseteq \ldots.$$

Let $\alpha_k(x,y)$ be L_{PA}-formulas and $[f_k] \in \mathbf{N}^*$ parameters such that $\alpha_k(x,[f_k])$ defines A_k (several parameters can be included in one, using the pairing function).

The finite intersection property now means simply that for all k

$$\mathbf{N}^* \models \exists x \alpha_k(x,[f_k])$$

which is equivalent to

$$\langle\langle \exists x \alpha_k(x,f_k) \rangle\rangle \in \mathcal{U}.$$

Let $[g_k]$ be a witness to the existential quantifier:

$$\langle\langle \alpha_k(g_k,f_k) \rangle\rangle \in \mathcal{U}.$$

What we want is a function $h \in \prod_\omega \mathbf{N}$ such that for all k

$$\langle\langle \alpha_k(h,f_k) \rangle\rangle \in \mathcal{U}.$$

We leave the task of devising a suitable function (using g_ks) as an exercise. (Hint: If you did not needed the non-principality of $\mathcal{U}$ something must be wrong.) □

Let us conclude this appendix by discussing the issue of the language of $\mathcal{M}$. In Section 1.1 we have chosen a much larger language L_{all} as the language of $\mathcal{M}$. This has the technical advantage that it contains all other languages that are needed in different constructions and we do not need to change the ambient model every time. The disadvantage is that L_{all} is huge; it has the cardinality of the continuum. In particular, it is not countable.

The $\aleph_1$-saturation that we invoked in Section 1.1 is a property of models that does imply properties (1) and (2) but for uncountable languages is stronger. On the other hand, the particular ultrapower with the index set ω we have presented here will give properties (1) and (2) for any language (but yields the $\aleph_1$-saturation only for countable ones).

This is why we proved properties (1) and (2) for $\mathbf{N}^*$ rather than the $\aleph_1$-saturation (which is also true as the language L_{PA} is finite): the argument from the proof of Theorem A.9 would work identically for any language (a subset of L_{all}). In any case, as mentioned in Section 1.1, these two properties (1) and (2) are essentially the only consequences of the assumption of the $\aleph_1$-saturation of $\mathcal{M}$ that we use in this book.

Standard notation, conventions and list of symbols

We have used some standard notation and conventions including:

- $[n] := \{1,\ldots,n\}$ while we identify n with $\{0,\ldots,n-1\}$.
- Notation $A(x_1,\ldots,x_k)$ implies that all free variables of the formula are among $x_1,\ldots,x_k$.

Conventions specific for the book include:

- Letters ϵ and δ always stand for a positive standard rational number.
- When we talk about positive infinitesimals we always denote them $1/t$ or similarly, for t a non-standard element of $\mathcal{M}$.

We point out a few facts that follow from the definitions but are mentioned here explicitly for clarity:

- Probabilities $\mathrm{Prob}_{\omega\in\Omega}[\ldots]$ are evaluated in $\mathcal{M}$ and hence their values are $\mathcal{M}$-rationals.
- If $\mu(b) = 1$ for $b = X/\mathcal{I}$, $X \in \mathcal{A}$, then it follows that the counting measure of X is infinitesimally close to 1 and so bounded from below by any $1-\epsilon$ ($\epsilon > 0$ is standard by the convention above).

We list now (in order of appearance) the specific symbols used in the book, pointing out the sections or chapters where they are defined.

Section 1.1:

- L_{all}: the language having symbols for all functions and relations on $\mathbf{N}$, the standard structure of natural numbers.
- $\mathcal{M}$ is an $\aleph_1$-saturated model of the theory of $\mathbf{N}$ in the language L_{all}.

Section 1.2:

- $\Omega \in \mathcal{M}$ is a sample space.
- $\mathcal{A} \in \mathcal{M}$ is the Boolean algebra of subsets Ω.
- $\mathcal{I} \subseteq \mathcal{A}$ is the ideal of sets having infinitesimal counting measure.
- $\mathcal{B}$ is the quotient algebra $\mathcal{A}/\mathcal{I}$.
- μ is the induced measure on $\mathcal{B}$ (Loeb's measure) with values in the standard reals $\mathbf{R}$.

Section 1.3:

- F denotes a family of random variables defined on Ω.
- $K(F)$: the Boolean-valued model resulting from F.
- $[\![\ldots]\!]$ is the Boolean truth evaluation of sentences.

Section 1.4:

- $\mathrm{Th}_\Gamma(L)$: the set of L-sentences in prefix class Γ that are true in $\mathbf{N}$.

Section 2.1:

- $d(a,b)$ is a metric on $\mathcal{B}$.

Section 3.2:

- F_{PV}: family of polynomial-time functions on $\{0,1\}^n$.

Chapter 4:

- A_{Sk}: a Skolemization of A
- $\mathrm{Sk}[A; f_1, \ldots, f_k]$: a conjunction of universal Skolem axioms

Section 5.1:

- L^2: second-order extension of language L
- L_{PA}: the language of PA
- $X < t$: bounding relation
- $\Sigma_0^{1,b}$, Σ_∞^b, $\Sigma_i^{1,b}$, $\Pi_i^{1,b}$, $s\Sigma_1^{1,b}$, $s\Pi_i^{1,b}$: various classes of bounded formulas

Section 5.2:

- $\mathcal{M}_n$: a cut in $\mathcal{M}$
- L_n and L_n^2: languages

Section 5.3:

- $K(F,G)$: a second-order structure

Chapter 6:

- $\mathrm{PA}^-, I\Delta_0, I\Delta_0(R), V_1^0$: various theories
- IND: the scheme of induction

Chapter 7:

- $\mathrm{dp}(T)$: the depth of a tree
- F_{rud}, G_{rud}, $K(F_{\mathrm{rud}}, G_{\mathrm{rud}})$, $L_n(F_{\mathrm{rud}}, G_{\mathrm{rud}})$: rudimentary families of random variables, the rudimentary structure and language

Section 8.1:

- $\langle\langle \ldots \rangle\rangle$ notation

Section 10.2:

- Ω_{tree}, F_{tree}, G_{tree} and $K(F_{\mathrm{tree}}, G_{\mathrm{tree}})$: the sample space and the families of the tree model

Section 12.1:

- $\mathrm{Par}(x, X, Y)$: a formula formalizing parity

Section 13.1:

- Q_2, $Q_2V_1^0$, $Q_2\Sigma_0^{1,b}$, $Q_2\Sigma_\infty^{1,b}$: parity quantifier, a resulting theory and classes of formulas

Section 13.2:

- 2-cover: a family of sets counting parity of sets in another family

Chapter 14:

- $\deg(T)$: the degree of an algebraic decision tree
- F_{alg}, G_{alg}, $K(F_{\mathrm{alg}}, G_{\mathrm{alg}})$: families of random variables and the algebraic tree model
- $\mathbf{F}_2^{\mathrm{low}}[x_1, \ldots, x_n]$: a ring of low-degree polynomials
- $\sum_i p_i \cdot i^*$: alternative formalism for elements of F_{alg}

Section 16.1:

- Closure(n): a formula expressing the existence of the transitive closure of a graph

Chapter 17:

- F, EF: Frege and Extended Frege systems
- $\oplus$, $F(\oplus)$: parity connective and a Frege system in the DeMorgan language with $\oplus$
- F_d and $F_d(\oplus)$: constant-depth subsystems

Chapter 18:

- $\mathrm{Fla}_d(x,Y)$, $\mathrm{Prf}_P(x,X,Y)$, $\mathrm{Sat}_d(x,Z,Y)$: $\Sigma_0^{1,b}$-formulas defining depth d formulas and the provability and the satisfiability predicates
- $\mathrm{Ref}_{P,d}(x,X,Y,Z)$, Ref_P: reflection principles

Chapter 19:

- $\mathrm{PHP}(x,R)$, PHP_m: first-order and propositional formulas formalizing the pigeonhole principle

Chapter 20:

- $K(F^0_{\mathrm{PHP}}, G^0_{\mathrm{PHP}})$ and Ω^0_{PHP}: a structure based on shallow PHP-trees and its sample space

Chapter 21:

- $K(F_{\mathrm{PHP}}, G_{\mathrm{PHP}})$ and Ω_{PHP}: a structure and its sample space

Section 22.1:

- $\mathbf{F}_2[x_{ij}]$, S, S^{low}, $\widetilde{S[\overline{r}]}$: rings
- $(\neg\mathrm{PHP}_n)$ a system of equations
- $Q_{i_1,i_2,j}$, $Q_{i;j_1,j_2}$, Q_i: polynomials

Section 22.3:

- $E_{i,\overline{g}}$, $\mathcal{E}$: an extension polynomial and their set

Section 22.4:

- $\hat{S}$: a ring
- $\hat{S}_t$, V_t: vector spaces

- C_t, Δ_t, T_t: vector space bases
- $x_c \preceq x_d$: ordering of terms
- Map, Map*: sets of finite maps

Section 22.5:

- $\sum_i p_i \cdot i^*$: an element notation
- $\hat{S}^{\text{low}}$: a ring

Chapter 23:

- L_{PV}: a language
- Σ_i^b, Π_i^b: classes of bounded formulas
- PV, S_2, T_2, S_2^i, T_2^i, U_1^1, V_1^1, BASIC: various theories
- PIND, LIND: schemes of polynomial induction and length induction

Chapter 24:

- WPHP(C,a): weak PHP for circuit C
- RSA$_k$: a particular sample space
- $\#FM\,(C,a)$: sharply bounded function minimization
- Sat(x,y): a PV formula defining the satisfaction relation
- Σ_1^b-LENGTH-MAX: a maximization principle

Chapter 25:

- Ω_0: a distribution on a sample space
- $\mathcal{A}_0$, $\mathcal{B}_0$: the resulting Boolean algebras
- $[\![\ldots]\!]_0$: the truth values in $\mathcal{B}_0$
- μ_0: the measure on $\mathcal{B}_0$
- $\equiv_c$: computational indistinguishability (weaker definition)

Chapter 26:

- Ω_{oracle}, $K(F_{\text{oracle}})$ and $K(F_{\text{oracle}}, G_{\text{oracle}})$: sample space and structures

Section 28.1:

- $||A||^n$: a propositional translation

Section 28.2:

- VPV: a theory
- Sat: a bounded formula

Section 29.1:

- $\tau(g)_b$: a propositional tautology
- Res_g^P: a set

Section 29.2:

- $\mathbf{tt}_{s,k}$: the truth-table function

Section 29.4:

- $\mathrm{NW}_{A,f}$: the Nisan–Wigderson generator
- $J_i(A)$: a set of columns in matrix A

Section 29.5:

- $g_{n,t}$: the PHP gadget generator

Section 30.1:

- CS: language Circuit Size

Section 30.2:

- $H_f(\ell)$: the hardness of a function

Section 30.3:

- $\mathsf{nw}_{n,c}$: a gadget generator

Section 31.1:

- $K(F_b)$, Ω_b, F_b: the local witness model and its sample space and family of random variables

Appendix:

- $\mathrm{Th}_{L_{\mathrm{PA}}}(\mathbf{N})$: the true arithmetic
- $\mathcal{U}$: an ultrafilter
- $\prod_\omega \mathbf{N}$, $\prod_\mathcal{U} \mathbf{N}$: a countable product and its reduction by an ultrafilter
- $\mathbf{N}^*$: a non-standard model

References

[1] M. Ajtai, Σ_1^1-formulae on finite structures, *Annals of Pure and Applied Logic*, **24** (1983), 1–48.

[2] M. Ajtai, The complexity of the pigeonhole principle, in *Proceedings of the IEEE 29th Annual Symposium on Foundation of Computer Science* (IEEE, 1988), pp. 346–55.

[3] M. Ajtai, Parity and the pigeonhole principle, in *Feasible Mathematics*, ed. S. R. Buss and P. J. Scott (Boston, MA: Birkhäuser, 1990), pp. 1–24.

[4] M. Ajtai, The independence of the modulo p counting principles, in *Proceedings of the 26th Annual ACM Symposium on Theory of Computing* (New York: ACM Press, 1994), pp. 402–11.

[5] M. Alekhnovich, E. Ben-Sasson, A. A. Razborov and A. Wigderson, Pseudorandom generators in propositional proof complexity, *Electronic Colloquium on Computational Complexity*, Rep. No. **23** (2000). Extended abstract in *Proceedings of the 41st Annual Symposium on Foundation of Computer Science* (IEEE, 2000), pp. 43–53.

[6] S. Arora and B. Barak, *Computational Complexity: A Modern Approach* (Cambridge, Cambridge University Press, 2009).

[7] P. Beame, R. Impagliazzo, J. Krajíček, T. Pitassi, P. Pudlák and A. Woods, Exponential lower bounds for the pigeonhole principle, in *Annual ACM Symposium on Theory of Computing* (New York: ACM Press, 1992), pp. 200–20.

[8] P. Beame, R. Impagliazzo, J. Krajíček, T. Pitassi and P. Pudlák, Lower bounds on Hilbert's Nullstellensatz and propositional proofs, *Proceedings of the London Mathematical Society*, **73**:3 (1996), 1–26.

[9] C. H. Bennett and J. Gill, Relative to a random oracle A, $P^A \neq NP^A \neq co{-}NP^A$ with Probability 1, *SIAM Journal of Computing*, **10**:1 (1981), 96–113.

[10] G. Boole, *The Mathematical Analysis of Logic* (Cambridge: Barclay and Macmillan, 1847).

[11] S. R. Buss, *Bounded Arithmetic* (Naples: Bibliopolis, 1986).

[12] S. R. Buss, Polynomial size proofs of the propositional pigeonhole principle, *Journal of Symbolic Logic*, **52** (1987), 916–27.

[13] S. R. Buss, Axiomatizations and conservation results for fragments of bounded arithmetic, in *Logic and Computation*, Contemporary Mathematics 106 (Providence, RI: American Mathematical Society, 1990), pp. 57–84.

[14] S. R. Buss, Some remarks on the lengths of propositional proofs, *Archive for Mathematical Logic*, **34** (1995), pp. 377–94.

[15] S. R. Buss, Relating the bounded arithmetic and polynomial time hierarchies, *Annals of Pure and Applied Logic*, **75** (1995), pp. 67–77.

[16] S. R. Buss, Lower bounds on Nullstellensatz proofs via designs, in *Proof Complexity and Feasible Arithmetics*, ed. S. Buss and P. Beame (Providence, RI: American Mathematical Society, 1998), pp. 59–71.

[17] S. R. Buss, and L. Hay, On truth-table reducibility to *SAT* and the difference hierarchy over *NP*, in *Proceedings of the IEEE Structure in Complexity Conference* IEEE, 1988), pp. 224–33.

[18] S. R. Buss, R. Impagliazzo, J. Krajíček, P. Pudlák, A. A. Razborov and J. Sgall, Proof complexity in algebraic systems and bounded depth Frege systems with modular counting, *Computational Complexity*, **6**:3 (1996/1997), pp. 256–98.

[19] S. R. Buss and J. Krajíček, An application of boolean complexity to separation problems in bounded arithmetic, *Proceedings of the London Mathematical Society*, **69**:3 (1994), pp. 1–21.

[20] S. R. Buss, J. Krajíček and G. Takeuti, On provably total functions in bounded arithmetic theories R^i_3, U^i_2 and V^i_2, in *Arithmetic, Proof Theory and Computational Complexity*, ed. P. Clote and J. Krajíček (Oxford: Oxford University Press, 1993), pp. 116–61.

[21] C. C. Chang and J. Keisler, *Model Theory*, Studies in Logic (Amsterdam: North-Holland 1977).

[22] V. Chvatal and E. Szemeredi, Many hard examples for resolution, *Journal of the ACM*, **35**:4 (1988), pp. 759–68.

[23] M. Chiari and J. Krajíček, Witnessing functions in bounded arithmetic and search problems, *Journal of the Symbolic Logic*, **63**:3 (1998), pp. 1095–115.

[24] M. Clegg, J. Edmonds and R. Impagliazzo, Using the Groebner basis algorithm to find proofs of unsatisfiability, in *Proceedings of the 28th Annual ACM Symposium on Theory of Computing* (New York: ACM Press, 1996), pp. 174–83.

[25] P. Clote and E. Kranakis, *Boolean Functions and Models of Computation* (Berlin: Springer-Verlag, 2002).

[26] A. Cobham, The intrinsic computational difficulty of functions, in *Proceedings of the 1964 Congress for Logic, Methodology and Philosophy of Science*, ed. Y. Bar-Hillel (Amsterdam: North-Holland, 1965), pp. 24–30.

[27] S. A. Cook, The complexity of theorem proving procedures, in *Proceedings of the* 3rd *Annual ACM Symposium on Theory of Computing* (New York: ACM Press, 1971), pp. 151–8.

[28] S. A. Cook, Feasibly constructive proofs and the propositional calculus, in *Proceedings of the* 7th *Annual ACM Symposium on Theory of Computing* (New York, ACM Press, 1975), pp. 83–97.

[29] S. A. Cook and J. Krajíček, Consequences of the provability of $NP \subseteq P/poly$, *Journal of Symbolic Logic*, **72**:4 (2007), pp. 1353–71.

[30] S. A. Cook and P. Nguyen, *Logical Foundations of Proof Complexity* (Cambridge: Cambridge University Press, 2009).

[31] S. A. Cook and R. A. Reckhow, The relative efficiency of propositional proof systems, *Journal of Symbolic Logic*, **44**:1 (1979), pp. 36–50.

[32] S. A. Cook and N. Thapen, The strength of replacement in weak arithmetic, preprint, *ACM Transactions on Computational Logic*, **7**:4 (2006).

[33] P. Erdös, Some remarks on the theory of graphs, *Bulletin of the American Mathematical Society*, **53** (1947), pp. 292–4.

[34] M. Furst, J. B. Saxe and M. Sipser, Parity, circuits and the polynomial-time hierarchy, *Mathematical Systems Theory*, **17** (1984), pp. 13–27.

[35] K. Gödel, a letter to John von Neumann from 1956, reprinted e.g. in *Arithmetic, Proof Theory and Computational Complexity*, ed. P. Clote and J. Krajíček (Oxford: Oxford University Press, 1993).

[36] O. Goldreich, *Foundations of Cryptography*, Vol. 1 (Cambridge: Cambridge University Press, 2001).

[37] O. Goldreich, Candidate one-way functions based on expander graphs, *Electronic Colloquium on Computational Complexity*, Rep. No. 90 (2000).

[38] P. Hájek and P. Pudlák, Metamathematics of first-order arithmetic, *Perspectives in Mathematical Logic* Springer-Verlag (Berlin: 1993).

[39] J. Hanika, Search problems and bounded arithmetic, PhD thesis, Charles University, Prague (2004).

[40] J. Hastad, Almost optimal lower bounds for small depth circuits, in *Randomness and Computation*, ed. S. Micali, Advances in Computing Research 5 (Stanford, CT: JAI Press, 1989), pp. 143–70.

[41] J. Hastad, R. Impagliazzo, L. Levin and M. Luby, A pseudorandom generator from any one-way function, *SIAM Journal on Computing*, **28** (1999), pp. 1364–96.

[42] E. A. Hirsch and D. Itsykson, An optimal heuristic randomized semidecision procedures, with application to proof complexity, preprint (2009).

[43] W. Hodges, *Model Theory*, Encyclopedia of Mathematics and its Applications 42 (Cambridge: Cambridge University Press, 1993).

[44] R. Impagliazzo, V. Kabanets, and A. Wigderson, In search of an easy witness: Exponential time vs. probabilistic polynomial time, *Journal of Computer and System Sciences*, **65**:4 (2002), pp. 672–94.

[45] R. Impagliazzo and M. Naor, Efficient cryptographic schemes provably as secure as subset sum, *Journal of Cryptology*, **9**:4 (1996), pp. 199–216.

[46] R. Impagliazzo, P. Pudlak and J. Sgall, Lower bounds for the polynomial calculus and the Groebner basis algorithm, *Computational Complexity*, **8**:2 (1999), pp. 127–44.

[47] R. Impagliazzo and N. Segerlind, Counting axioms do not polynomially simulate counting gates, in *Proceedings of the IEEE* 42nd *Annual Symposium on Foundation of Computer Science* (IEEE, 2001), pp. 200–9.

[48] R. Impagliazzo and A. Wigderson, P = BPP unless E has sub-exponential circuits: derandomizing the XOR lemma, in *Proceedings of the 29th Annual ACM Symposium on Theory of Computing* (New York: ACM Press, 1997), pp. 220–9.

[49] T. Jech, *Set Theory* (Berlin: Springer-Verlag, 2003).

[50] E. Jeřábek, Dual weak pigeonhole principle, Boolean complexity, and derandomization, *Annals of Pure and Applied Logic*, **129** (2004), pp. 1–37.

[51] V. Kabanets and J.-Y. Cai, Circuit minimization problem, Extended abstract in *Proceedings of the 32nd Annual ACM Symposium on Theory of Computing* (New York: ACM Press, 2000) pp. 73–9. Also: Technical Report TR99–045, Electronic Colloquium on Computational Complexity.

[52] R. Kaye, *Models of Peano Arithmetic* (Oxford: Oxford Science, 1991).
[53] J. Krajíček, A fundamental problem of mathematical logic, *Annals of the Kurt Gödel Society*, Collegium Logicum, Vol. 2 (Berlin: Springer-Verlag, 1996), pp. 56–64.
[54] J. Krajíček, Fragments of bounded arithmetic and bounded query classes, *Transactions of the American Mathematical Society*, **338**:2 (1993), pp. 587–98.
[55] J. Krajíček, On Frege and Extended Frege proof systems, in *Feasible Mathematics II*, ed. P. Clote and J. Remmel (Boston, MA: Birkhauser, 1995), pp. 284–319.
[56] J. Krajíček, *Bounded Arithmetic, Propositional Logic, and Complexity Theory*, Encyclopedia of Mathematics and Its Applications 60 (Cambridge: Cambridge University Press, 1995).
[57] J. Krajíček, On methods for proving lower bounds in propositional logic, in *Logic and Scientific Methods: Volume One of the Proceedings of the Tenth International Congress of Logic, Methodology and Philosophy of Science, Florence*, August 1995, Synthese Library, 259, ed. M. L. Chiara, K. Doets, D. Mundici and J. F. A. K. Benthem (Dordrecht: Kluwer Academic, 1997), pp. 69–83.
[58] J. Krajíček, Lower bounds for a proof system with an exponential speed-up over constant-depth Frege systems and over polynomial calculus, in *Mathematical Foundations of Computer Science: 22nd International Symposium Bratislava, August 1997*, ed. I. Prívara and P. Růžička, Lecture Notes in Computer Science 1295 (Berlin: Springer-Verlag, 1997), pp. 85–90.
[59] J. Krajíček, On the degree of ideal membership proofs from uniform families of polynomials over a finite field, *Illinois Journal of Mathematics*, **45**:1 (2001), pp. 41–73.
[60] J. Krajíček, Extensions of models of *PV*, in *Logic Colloquium'95*, ed. J. A. Makowsky and E. V. Ravve, Lecture Notes in Logic 11 (Berlin: Springer-Verlag, 1998), pp. 104–114.
[61] J. Krajíček, On the weak pigeonhole principle, *Fundamenta Mathematicae*, **170**:1 (2001), pp. 123–40.
[62] J. Krajíček, Tautologies from pseudo-random generators, *Bulletin of Symbolic Logic*, **7**:2 (2001), pp. 197–212.
[63] J. Krajíček, Dual weak pigeonhole principle, pseudo-surjective functions, and provability of circuit lower bounds, *Journal of Symbolic Logic*, **69**:1 (2004), pp. 265–86.
[64] J. Krajíček, Diagonalization in proof complexity, *Fundamenta Mathematicae*, **182** (2004), pp. 181–92.
[65] J. Krajíček, Implicit proofs, *Journal of Symbolic Logic*, **69**:2 (2004), 387–97.
[66] J. Krajíček, Structured pigeonhole principle, search problems and hard tautologies, *Journal of Symbolic Logic*, **70**:2 (2005), pp. 619–30.
[67] J. Krajíček, Hardness assumptions in the foundations of theoretical computer science, *Archive for Mathematical Logic*, **44**:6 (2005), pp. 667–75.
[68] J. Krajíček, Proof complexity, in *European Congress of Mathematics, Stockholm, June 27–July 2, 2004*, ed. A. Laptev (Zurich: European Mathematical Society, 2005), pp. 221–31.
[69] J. Krajíček, Substitutions into propositional tautologies, *Information Processing Letters*, **101**:4 (2007), pp. 163–7.

[70] J. Krajíček, A proof complexity generator, in *Logic, Methodology and Philosophy of Science: Proceedings of the Thirteenth International Congress*, ed. C. Glymour, W. Wang and D. Westerstahl (London: King's College Publications, 2009), pp. 185–90.

[71] J. Krajíček, A form of feasible interpolation for constant depth Frege systems, *Journal of Symbolic Logic*, **75**:2 (2010), pp. 774–84.

[72] J. Krajíček, On the proof complexity of the Nisan-Wigderson generator based on a hard NP∩coNP function, submitted (preprint March 2010). Preliminary version in *Electronic Colloquium on Computational Complexity*, Rep. No. 54 (2010).

[73] J. Krajíček and P. Pudlák, Propositional proof systems, the consistency of first order theories and the complexity of computations, *Journal of Symbolic Logic*, **54**:3 (1989), pp. 1063–79.

[74] J. Krajíček and P. Pudlák, Quantified propositional calculi and fragments of bounded arithmetic, *Zeitschr. f. Mathematikal Logik u. Grundlagen d. Mathematik*, **36**:1 (1990), pp. 29–46.

[75] J. Krajíček and P. Pudlák, Propositional provability in models of weak arithmetic, in *Computer Science Logic '89*, ed. E. Börger, H. Kleine Bünning and M. M. Richter, Lecture Notes in Computer Science 440 (Berlin: Springer-Verlag, 1990), pp. 193–210.

[76] J. Krajíček and P. Pudlák, Some consequences of cryptographical conjectures for S^1_2 and *EF*, *Information and Computation*, **140**:1 (1998), pp. 82–94.

[77] J. Krajíček, P. Pudlák and J. Sgall, Interactive computations of optimal solutions, in *Mathematical Foundations of Computer Science* (B. Bystrica, August '90), ed. B. Rovan, Lecture Notes in Computer Science 452 (Berlin: Springer-Verlag, 1990), pp. 48–60.

[78] J. Krajíček, P. Pudlák and G. Takeuti, Bounded arithmetic and the polynomial hierarchy, *Annals of Pure and Applied Logic*, **52** (1991), pp. 143–53.

[79] J. Krajíček, P. Pudlák and A. Woods, An exponential lower bound to the size of bounded depth frege proofs of the pigeonhole principle, *Random Structures and Algorithms*, **7**:1 (1995), pp. 15–39.

[80] P. A. Loeb, Conversion from nonstandard to standard measure spaces and applications in probability theory, *Transactions of the American Mathematical society*, **211** (1975), pp. 113–22.

[81] A. Maciel and T. Pitassi, Towards lower bounds for bounded-depth Frege proofs with modular connectives, in *Proof Complexity and Feasible Arithmetics*, ed. P. Beame and S. Buss, DIMACS Series in Discrete Mathematics and Theoretical Computer Science 39 (Providence, RI: American Mathematical Society, 1998), pp. 195–227.

[82] R. Mansfield, The theory of boolean ultrapowers, *Annals of Mathematical Logic*, **2**:3 (1971), pp. 297–323.

[83] D. Marker, *Model Theory: An Introduction*, Graduate Texts in Mathematics 217 (Berlin: Springer, 2002).

[84] N. Nisan and A. Wigderson, Hardness vs. randomness, *Journal of Computer and System Sciences*, **49** (1994), pp. 149–67.

[85] R. Parikh, Existence and feasibility in arithmetic, *Journal of Symbolic Logic*, **36** (1971), pp. 494–508.

[86] J. Paris and A. J. Wilkie, Counting problems in bounded arithmetic, in *Methods in Mathematical Logic*, Lecture Notes in Mathematics 1130 (Berlin: Springer-Verlag, 1985), pp. 317–40.
[87] J. Pich, Nisan-Wigderson generators in proof systems with forms of interpolation, submitted (preprint 2010). Preliminary version in *Electronic Colloquium on Computational Complexity*, Rep. No. 46, (2010).
[88] T. Pitassi, P. Beame and R. Impagliazzo, Exponential lower bounds for the pigeonhole principle, *Computational Complexity*, **3** (1993), pp. 97–308.
[89] P. Pudlák, The lengths of proofs, in *Handbook of Proof Theory*, ed. S. R. Buss (Amsterdam: Elsevier, 1998), pp. 547–637.
[90] H. Rasiowa and R. Sikorski, Algebraic treatment of the notion of satisfiability, *Fundamenta Mathematicae*, **40** (1953), pp. 62–5.
[91] H. Rasiowa and R. Sikorski, *The Mathematics of Metamathematics* (Warsaw: PWN, 1963).
[92] A. A. Razborov, Lower bounds on the size of bounded depth networks over a complete basis with logical addition, *Matem. Zametki*, **41**:4 (1987), pp. 598–607.
[93] A. A. Razborov, Unprovability of lower bounds on the circuit size in certain fragments of bounded arithmetic, *Izvestiya of the Rossiiskoi Akademii Nauk*, **59**:1 (1995), pp. 201–24.
[94] A. A. Razborov, Lower Bounds for the Polynomial Calculus, *Computational Complexity*, **7**:4 (1998), pp. 291–324.
[95] A. A. Razborov, Resolution lower bounds for perfect matching principles, in *Proceedings of the 17th IEEE Conference on Computational Complexity* (IEEE, 2002), pp. 29–38.
[96] A. A. Razborov, Pseudorandom generators hard for k-DNF resolution and polynomial calculus resolution, preprint (May 2003).
[97] A. A. Razborov and S. Rudich, Natural proofs, *Journal of Computer and System Sciences*, **55**:1 (1997), pp. 24–35.
[98] S. Riis, Independence in bounded arithmetic, DPhil thesis, Oxford University, (1993).
[99] S. Rudich, Super-bits, demi-bits, and $\tilde{NP}/qpoly$-natural proofs, in *Proceedings of the 1st International Symp. on Randomization and Approximation Techniques in Computer Science*, Lecture Notes in Computer Science 1269 (Berlin: Springer-Verlag, 1997), pp. 85–93.
[100] D. Scott, A proof of the independence of the continuum hypothesis, *Mathematical Systems Theory*, **1** (1967), pp. 89–111.
[101] R. Smolensky, Algebraic methods in the theory of lower bounds for Boolean circuit complexity, in *Proceedings of the 19th Annual ACM Symposium on the Theory of Computing* (New York: ACM Press, 1987), pp. 77–82.
[102] G. Takeuti, *Two Applications of Logic to Mathematics* (Princeton, NJ: Princeton University Press, 1978).
[103] G. Takeuti and M. Yasumoto, Forcing in bounded arithmetic, in *Gödel'96: Logical Foundations of Mathematics, Computer Science, and Physics*, ed. P. Hájek, Lecture Notes in Logic 6 (Natick, MA: A. K. Peters, 1996), pp. 120–38.
[104] G. Takeuti and M. Yasumoto, Forcing in bounded arithmetic II, *Journal of Symbolic Logic*, **63** (1998), pp. 860–8.

[105] G. Takeuti and W. Zaring, *Axiomatic Set Theory*, Graduate Texts in Mathematics 8 (Berlin: Springer, 1973).

[106] E. Viola, Correlation bounds for polynomials over $\{0,1\}$, *ACM SIGACT News*, **40**:1 (2009).

[107] K. W. Wagner, Bounded query classes, *SIAM Journal of Computing*, **19**:5 (1990), pp. 833–46.

[108] A. Yao, Separating the polynomial-time hierarchy by oracles, in *Proceedings of the 26th Annual IEEE Symposium on Foundations of Computer Science* (IEEE, 1985), pp. 1–10.

[109] D. Zambella, Notes on polynomially bounded arithmetic, *Journal of Symbolic Logic*, **61**:3 (1996), pp. 942–66.

Name index

Subject index

Printed in the United States
by Baker & Taylor Publisher Services